Night Sky

SECOND EDITION

Nicholas Nigro

Guilford, Connecticut

FALCON®

An imprint of The Rowman & Littlefield Publishing Group, Inc.
4501 Forbes Blvd., Ste. 200
Lanham, MD 20706
www.rowman.com
Falcon and FalconGuides are registered trademarks and Make Adventure
Your Story is a trademark of The Rowman & Littlefield Publishing Group,
Inc.

Distributed by NATIONAL BOOK NETWORK

Copyright © 2012, 2021 The Rowman & Littlefield Publishing Group, Inc.

All rights reserved. No part of this book may be reproduced in any form
or by any electronic or mechanical means, including information storage
and retrieval systems, without written permission from the publisher, except by a reviewer who may quote passages in a review.

British Library Cataloguing in Publication Information available

Library of Congress Cataloging-in-Publication Data available

ISBN 978-1-4930-5543-2 (paper : alk. paper)
ISBN 978-1-4930-5544-9 (electronic)

∞™ The paper used in this publication meets the minimum requirements
of American National Standard for Information Sciences—Permanence of
Paper for Printed Library Materials, ANSI/NISO Z39.48-1992.

Contents

Introduction

Astronomy is the only branch of science where amateurs routinely make considerable contributions and pioneering discoveries. Look upon this field guide as a colorful welcome mat to the art of stargazing—a handy primer introducing you to a hobby with unlimited possibilities. The greater your understanding of the cosmic players on the nighttime stage, the more you will appreciate the incredible vastness, breathtaking beauty, and enduring mysteries of our solar system and the universe beyond.

Amateur astronomers are not bound by any particular playbook—quite the contrary. You can gaze into the cosmos with your own two eyes and nothing else, or you can employ the magnifying muscle of a pair of binoculars or, better still, a telescope. You can survey the familiar or the unfamiliar. You can investigate the moon, a stellar stone's throw away, or the more distant and diverse planets that revolve around our sun. You can pinpoint individual stars and call on space neighborhoods known as constellations. You can also venture very far afield and hunt down deep-sky objects that are hundreds, thousands, and even millions of light-years away.

As a committed "backyard astronomer," a term sometimes applied to starwatchers, you'll fast recognize that the night sky's the limit. And this field guide will furnish you with indispensable information and practical tips on maximizing the results of all your astronomical adventures. It'll show you exactly where to look in the busy Northern Hemisphere night skies, what to look for, and explain what you are seeing, too. In addition, by dispensing concise and practical counsel on such vital matters as optimal weather conditions, preferred locations, and, of course, utilization of the various tools of the trade, this book will assist you in adeptly navigating to and fro the vast and dynamic nighttime skies.

Something extraordinary happens to individuals of all ages, in all parts of the world, and from all walks of life when they cast their sights and imaginations beyond the cozy confines of planet Earth. The magnificent and mystifying celestial ether calls them back again and again. It is this unquenchable thirst to uncover more and more in the far reaches of outer space that unites stargazers the world over.

Contemplate this as you cast your eyes and imagination heavenward and gaze into the unfathomable ether of outer space: Stars are born and die every day. There are vast and varied estimates of their numbers in the observable universe—as many as one billion trillion and 400 billion in our galaxy alone—because so much is still unknowable. When considering the disparity in age of the universe and our home planet, light from the farther reaches of the former has yet to reach us. The night sky is now and will always be a work in progress, at once reassuringly familiar and spectacularly mysterious.

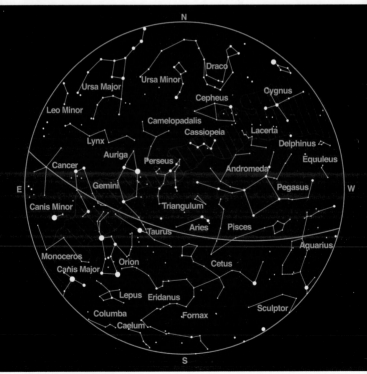

Illustration by David Cole Wheeler

Stargazing trifecta: clear skies, wind free, and away from artificial light sources
© Thomas Nigro

CHAPTER 1 Stargazing ABCs

The Right Weather Conditions

Maximizing the outcomes of your astronomical adventures requires a helping hand from Mother Nature. Before any and all stargazing endeavors, it's a good idea to pay special heed to weather forecasts. However, bear in mind that optimal conditions for surveying the nighttime skies involve more than the absence of storm clouds. And unless the moon is your celestial prey, a thin crescent or, better still, no visible moon at all is preferable.

For starters, excessive moonlight diminishes the all-important darkness that will furnish you with the most favorable portal into the celestial beyond. Even a stiff breeze will impede your viewing pleasure. Winds churn up the atmosphere at the surface, lower visibility, and make the less-bright night-sky objects more difficult, and often impossible, to locate. The most potent one-two starwatcher's punch is a simultaneously clear and tranquil night.

© Changhua Ji | Dreamstime.com

Courtesy of their generally agreeable weather conditions and velvety dark night skies, some favorite stargazing hot spots include the American Southwest's Sonoran Desert, Bryce Canyon in Utah, Grand Canyon in Arizona, and Yosemite and Death Valley in California. Utah's Natural Bridges National Monument has actually been christened an "International Dark Sky Park." There are, in fact, astronomical hotspots in every region of the country, including Cherry Springs State Park in Coudersport, Pennsylvania; Dry Tortugas National Park in the Florida Keys; Denali National Park in Interior Alaska; and elevated and isolated Maunakea in Hawaii.

Since most amateur astronomy jaunts occur during the summer months—when, in fact, the night skies put on some of their most impressive performances—there is often a matter of relative humidity to contend with. Suffice it to say that less is better where humidity is concerned.

Excessive moisture in the air sullies the best possible picture window into outer space. Often humidity levels—just like wind speeds—are overlooked in the planning stages, but they shouldn't be. Many deep-sky objects, for instance, are observable only under ideal conditions. A very humid night or an exceptionally breezy evening will take them out of the celestial snapshot altogether.

Something else to ponder vis-à-vis general weather conditions is the season of the year. The crisp, drier, less-polluted air found during the winter months—along with the accessible positioning of many prominent constellations, including Orion and Canis Major—deliver impressive sky shows. Yet the colder temperatures, coupled with the inherent difficulties in remaining in the great outdoors for extended periods of time, make wintertime stargazing decidedly less popular than its summertime cousin. Dress for the season, but don't let winter's night skies go unnoticed.

Optimum Locations

While the science of astronomy is sometimes complex and bewildering, the stargazing hobby is rooted in simplicity. True, not all skywatching locales are created equal, but you can in fact peruse the nighttime skies from just about anywhere. Job one

Dark locales like national parks and beaches supply optimal stargazing possibilities.
© Thomas Nigro

is to identify a safe and stable spot in which to canvass the cosmic beyond. Rooftops and other high elevations may supply you with a grand spectacle, but they are dangerous places to be when looking heavenward.

If you can glance north and south, then east and west, and spy the celestial sphere in total, you have surmounted a considerable stargazing hurdle. Naturally, avoid setting up shop near tall trees. Buildings, or structures of any kind, interfere with the astronomical ideal—a sweeping panorama.

Worth noting, too, is that close proximity to the bright lights of big cities is bad for the business at hand. Excessive lighting cast into the nighttime skies removes countless celestial objects from your viewfinder. Merely being in the vicinity of streetlights or ordinary car lights is enough to appreciably reduce outer-space sightings.

In antiquity, many more cosmic players were visible on the nighttime stage than can be seen today. Man-made lighting and pollutants are the chief culprits. Skies were a whole lot darker before a thing called electricity, and less despoiled with soot and smog before the advent of industrialization.

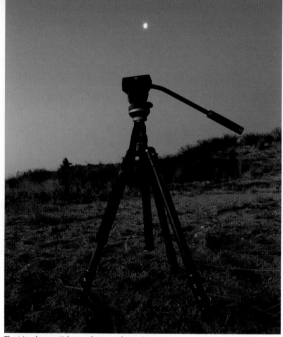
The tripod: essential astrophotography equipment
© Thomas Nigro

A general stargazing rule of thumb is therefore this: Rural is better than urban; clean air is better than foul air. Not surprisingly, national parks are very popular places for amateur astronomers to assemble. They furnish premium views into the night skies in safe, spacious topography. If feasible, inspect the night skies in a known park, or in an open field as far away as humanly possible from synthetic light sources—the bane of starwatchers everywhere.

And if you are stargazing in the cold climes of wintertime, make absolutely certain that you are suitably attired for the occasion. Perusing the night skies involves a fair share of immobility. Dress with this physical level of activity—or inactivity, as it were—in mind. Checking out the stars on high is a nighttime event with frequently falling temperatures, even in summertime and in desert locations as well.

© Diego Cervo | Dreamstime.com

Basic Gear

Before investing in a pair of binoculars or telescope, many pro-fessional astronomers recommend that beginners employ naked-eye observation of the night sky. It's critical to first ascertain where key outer-space markers are, such as the North Star and the Big Dipper. The unaided eye enables you to effortlessly traverse the celestial sphere east to west and north to south.

Binoculars are nevertheless invaluable gear for this hobby. They are less involved instruments than telescopes and thus well suited for amateurs. Stargazers typically utilize binoculars with 7 x 50 or 10 x 50 magnification. More advanced models feature 15 x 70, 20 x 80, and 25 x 100. The wider the field of view on the bin-oculars, the more celestial light-ing it will harvest, which is what you want.

It's also a prudent idea to sky-watch with an LED red flashlight. This type of light does not ham-per night vision. Contrarily, any white-lighting sources—even

© Ktphotog | Dreamstime.com

5

© Photohare | Dreamstime.com

the negligible flash of a lighter or a single match—will set your night vision back. Your pupils require full dilation, which takes approximately thirty to forty-five minutes, for optimal night vision and the most intimate look-see into outer space.

Telescopes

The telescope is practically synonymous with the science of astronomy. While it's not necessary to survey the night skies with one, the right telescope will enable you to see space objects more clearly and with more contrast than you would otherwise with the unaided eye or binoculars. While the latter are extremely beneficial to the hobby, they do not enhance celestial targets with intricate details. Telescopes do. They also permit you to locate things that are not visible without their incomparable magnifying prowess.

There are numerous styles of telescopes that range far and wide in size, appearance, and price. Nevertheless, all telescopes are classified as one of three varieties: refractors, reflectors, or catadioptrics.

A refractor telescope employs what are called objective lenses, which in essence "twist" light sources as they accumulate at their substantial heads. Refractors are excellent telescopes for general night-sky viewings but are not well suited for hunting down deep-sky objects in the farther regions of space.

Small 2.4-inch (60-millimeter) or 3-inch (80- to 90-millimeter) refractors are excellent for beginners and will provide you with ample observational grist.

Sometimes referred to as "Newtonian telescopes," reflector telescopes—unlike refractors—use mirrors to accumulate celestial light sources. In so doing, reflectors are renowned for their panoramic viewing fields, considerably wider than what their refractor competitors offer.

Courtesy of these more expansive vistas of outer space, reflectors are thus enormously effective at locating dim deep-sky objects, including faraway galaxies and gaseous, dust-laden nebulae. The application of mirrors, instead of more costly lenses, translates into a less expensive product, too. Indeed, reflectors are typically the cheapest telescopes on the market. There are many varieties of this telescope with widely varying capacities. The Hubble Space Telescope is actually a sophisticated version of a Newtonian telescope.

In addition to their affordability, reflector telescopes are generally seen as the most comfortable to employ in the field, too, with their eyepieces at convenient heights. For zeroing in on planets and pinpointing distant galaxies, amateurs often gravitate toward Dobsonian telescopes, a type of reflector. In conspicuous contrast with many rival models, Dobsonian telescopes are at once easy to mount and remarkably durable in the rough and tumble of the great outdoors.

There is a third classification of telescope known as the catadioptric—an amalgam that combines elements of both refractors and reflectors. Catadioptric telescopes are more popularly known as "Schmidt-Cassegrain," and their derivatives, Maksutov-Cassegrain, Classical Cassegrain, etc. Since they are not exactly central casting's image of a telescope, you might consider one a bit peculiar-looking.

While blending technologies that effectively work for both refractors and reflectors, catadioptrics also tout various computer-enhanced models. They are typically less expensive than refractors but more expensive than reflectors. As something of a telescopic hybrid, catadioptrics are

© Raymond Kasprzak | Dreamstime.com

very portable and offer a mother lode of accessories to enhance one's viewing experience.

Before purchasing any genus of telescope, there are three essential aspects to consider: light-gathering capacity, quality of the optics, and ease of mounting. Some popular manufacturers of telescopes include Meade, Celestron, and Orion. Svbony makes a popular refractor telescope, which is ideal for beginners and affordable as well. It's always a sensible idea, too, to avoid the willy-nilly purchasing of telescopes and all other astral accessories. Specialized shops with knowledgeable staff are the best places to discover the right astronomical equipment for you and your unique station in the hobby.

Astrophotography

Should you be interested in capturing the goings-on in the night sky with a camera, modern technology can lend a helping hand. Astrophotography, as it is known, has now completely moved beyond film to the digital realm, which supplies one and all with more room for experimentation. In other words, you can take many, many pictures without worrying about misfires and expensive film.

As a novice in this potentially rewarding endeavor, photographing the night sky and so much of what it has to offer—including the Milky Way, planets, the moon, stars, and celestial pageantry like comets and meteor showers—is eminently doable. For starters, it's imperative that you grasp the fundamentals of photography and the importance of your camera's ISO setting, which lightens and darkens images. The key, too, to establishing depth and clarity in astrophotography—of faraway celestial bodies and snippets of the sprawling night sky—is the long exposure. Today's cameras with their highly advanced sensors can nobly assist with the job at hand.

A basic DSLR (digital single-lens reflex) camera and a tripod—critical for stability—is essentially all you need to get started. This equipment, of course, and a clear, relatively wind-free night. As you progress in the province of astrophotography, more advanced options will present themselves, including the use of a telescope for capturing deep-sky objects and happenings.

CHAPTER 2

Nearby Celestial Sights

The Moon

Natural satellites, or moons, are celestial objects that orbit a planet or smaller space body like an asteroid. Our moon is among the largest natural satellites in relation to its parent planet's dimensions. At approximately one-quarter the size of Earth, and on average 238,000 miles away, it is understandably a prime stargazing target.

When the moon is up, it is the brightest object in the night sky. There are no close runner-ups, either. When hugging the horizon, it appears much larger than normal to our probing eyes. However, the moon is actually a bit smaller and farther away from our planet at these times—a celestially orchestrated optical illusion, if you will—than when it is high in the sky. And while the

From our Earthly perspective: Moon size is often a spectacular optical illusion.
© Thomas Nigro

moon shines brightly in the night—reflecting sunlight—it is, as solar system bodies go, a rather poor reflector of light.

When inspecting the moon's craggy and barren contours, never lose sight of the fact that you are seeing only its so-called "near side" and never its "far side." The moon orbits synchronously with one hemisphere always facing Earth.

Vivid dividing line, the terminator, distinguishes the moon's light and dark hemispheres.
© Thomas Nigro

The moon has long been night-sky quarry, a particular favorite of beginners in the hobby fascinated by this big, bright, and accessible celestial neighbor of ours. If the moon is the object of your affections, its well-lit persona works to your benefit—even though an astronomer's most trusted friend is darkness.

The moon graces the night sky in a perpetual series of phases. There are eight in total throughout each lunar cycle, which lasts the better part of a calendar month: new moon, waxing crescent, first quarter, waxing gibbous, full moon, waning gibbous, last quarter, and waning crescent.

While there are no "best" moon observational times per se, there is something of an astronomical consensus that several days into the first-quarter phase furnishes one with premier views of the lunar environment. Indeed, both quarter-moon phases showcase sunlight emanating from either side of the moon, which permits sunrises and sunsets—from our earthly perspective—to cut a swath right through its midsection. Such indirect lighting enables the moon's haunting intricacies to reveal themselves—to quite literally shine.

Planets

A planet is a celestial object orbiting a star. Earth orbits the sun, which is classified as a yellow dwarf star, and so do seven other planets: Mercury, Venus, Mars, Jupiter, Saturn, Uranus, and Neptune. Pluto, which once was accorded planet status, has been downgraded to "dwarf planet" by the International Astronomical Union (IAU). The IAU has established three criteria for planetary designation. A planet must orbit the sun. A planet must maintain sufficient mass to, in layman's terms, take on a more-or-less round shape. And finally—and this is where Pluto falls short—a planet must have "cleared the neighborhood" of its outer space

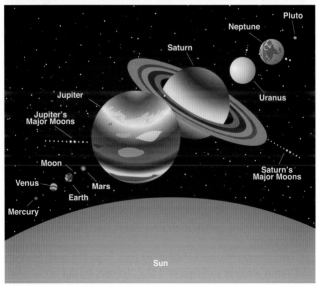

Illustration by David Cole Wheeler

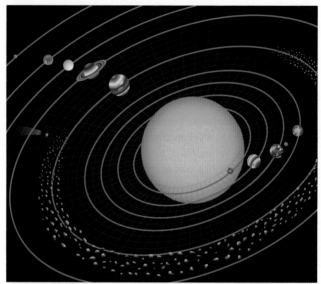

Illustration by David Cole Wheeler

orbit. In Pluto's orbit, for instance, there are celestial bodies of similar mass. With the exceptions of Neptune and Pluto, each one of the planets can be seen with the unaided eye at various points during the year.

When the moon is up, it is the brightest object in the night sky bar none. Venus is the brightest planet and, depending on its orbital position, often visible in the early mornings or early evenings. Look for Venus fairly low in the sky, although it is never visible throughout an entire night. Contrarily, Jupiter is discernible in different areas of the sky all night long. Trailing only the moon and Venus, Jupiter is the next-brightest night-sky object.

The planets nearest the sun are Mercury, Venus, Earth, and Mars. The planets farthest from the sun are Jupiter, Saturn, Uranus, and Neptune.

We inhabit the so-called third rock from the sun. Mercury and Venus are one and two, and Mars follows on our home planet's heels. Among these four inner planets—also known as the

Courtesy of Lunar and Planetary Institute

"terrestrials" because of their rocky and metallic compositions—Earth is biggest in size.

The outer planets, Jupiter, Saturn, Uranus, and Neptune—dubbed the "gas giants"—paint an entirely different portrait in both mass and makeup. Indeed, the inner planets are significantly smaller than the outer planets. Jupiter alone constitutes more than two and a half times the mass of all the other planets in our solar system combined. Mercury, the nearest planet to the sun, is the smallest of the stellar troupe and less than half of Earth's dimensions.

All of the planets in our solar system, big and small, are nonetheless tiny specks in the Milky Way Galaxy, estimated at 100,000 light-years in diameter. And the Milky Way is just one of billions of galaxies beyond its celestial borders. With the aid of telescopes, many galaxies are accessible for observation, with some revealing conspicuous details like spiral arms and radiant nuclei.

Courtesy of being both the nearest planet to the sun and the smallest in size, Mercury is not readily spotted in the night sky. Whereas Earth is on average 93 million miles away from the sun, Mercury, by comparison, is just 36 million miles away. Mercury orbits the sun approximately once every eighty-eight Earth

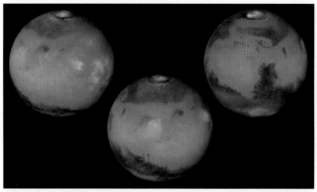

Courtesy of NASA/ David Crisp and the WFPC2 Science Team (Jet Propulsion Laboratory/California Institute of Technology)

days. The planet also sports what scientists deem a primitive atmosphere—razor-thin and therefore incapable of intercepting and disintegrating meteoroids. This flimsy defense accounts for Mercury's heavily cratered, moon-like surface.

Venus, second-closest planet to the sun, is just a trace smaller in overall diameter to Earth and unusually bright in the night sky. When it is nearing Earth in its orbit, it is detectable during the early evening hours. When Venus is moving away from Earth, it is perceptible in the early morning hours; hence, its dual nicknames of "Evening Star" and "Morning Star."

As an amateur astronomer exploring our solar system's myriad planets, you'll quickly come to appreciate their astonishing diversity. In utter contrast with gray, desolate Mercury, for instance, Venus's surface—although shrouded by dense clouds— is noticeably shaped by extensive volcanic activity.

Mars is certainly among the most famous of planets. Popular culture has seen to that. It is the only planet that human beings could conceivably visit at some future date. With its pinkish-red hues on the celestial palette, Mars is often referred to as the "Red Planet." This curious color scheme is the consequence of widespread iron oxide on its surface.

Mars's atmosphere is extremely slender and very clear. Clouds can thus be spotted now and again by telescopic observers, as can occasional dust storms initiated by solar heating of said atmosphere and the subsequent rapid movement of air, which kicks up phenomenal amounts of dust. At their most violent, these storms have entirely enveloped the planet.

Depending on its orbiting position, even the most basic of telescopes will afford you an insider's view of Mars's diverse and riveting surface. Tracking down the planet in the night sky is highly recommended. Thanks to its slight atmosphere, which is only 1 percent as dense as Earth's atmosphere at sea level, a captivating stargazing visual is always possible. Appearing as a reddish star, Mars is unmistakable in the celestial lineup.

Space Phenomena

As you read these words, meteoroids are battering our planet's upper atmosphere. Most of them are no bigger than a pebble and

© Tudorica Alexandru | Dreamstime.com

© Wikipedia Commons/photo by Senior Airman Joshua Strang

some even as small as a grain of sand. The vast majority of meteoroids burn up in Earth's protective buffer and cause streaks of light called "meteors."

Meteors regularly appear in the night sky as wayward stars knocked off their fixed celestial perches. Although sometimes called "shooting" or "falling" stars, they are actually pieces of rocky or metallic space matter—far removed from massive and burning stars in the throes of nuclear fusion. Some meteors don't fully disintegrate in the atmosphere's vise, and when they touch down on Earth's surface, they are officially and forevermore "meteorites."

Meteors can be spotted during all four seasons of the year, although there are specific times when annual meteor showers occur. These space showers arise when our planet's conventional orbit intersects with a comet's or asteroid's lingering trail of debris at around the same time every year. Both meteors and meteor showers can be very satisfactorily observed with the naked eye.

Earth's immediate and densest atmospheric layer—the troposphere—extends up from the planet's surface some 10 miles. It is here where the ever-changing events that we call "weather" transpire. In the farthest region of Earth's atmosphere, the thermosphere, a decidedly different type of weather occurs.

Christened "space weather," this phenomenon reveals itself in a layer of our atmosphere that is reedy-thin. Highly volatile energy particles—the upshot of unremitting solar radiation—interact with Earth's near-space boundary. Atoms and molecules therein react to this incessant barrage from the interplanetary medium and become "excited." They return to their non-excited ground states—a not inconsiderable feat—by emitting photons, which spawn multihued and undulating curtains of light.

Often visible in the upper atmosphere near the Polar Regions, these mesmerizing light shows are known as "aurorae" (singular: aurora). In the Northern Hemisphere they are called the Aurora Borealis or the Northern Lights. To increase your chances of seeing this spectacular phenomenon, it is imperative that you venture to the highest northern latitudes. Northern Alaska and surrounding areas are prime viewing locales for the Aurora Borealis. They have, however, been visible in places ranging far and wide from Northern Maine to the Upper Peninsula of Michigan to the Dakotas. Typically, the best viewings occur—after dark—from September through March.

© Wikipedia Commons | ESO/Y. Beletsky

CHAPTER 3 Distant Space Attractions

The Milky Way

Quite often the most impressive visual on the nighttime stage is the Milky Way, a spiral galaxy and the celestial address of our solar system. The Milky Way appears as a creamy but very vivid swath of light arcing across the celestial sphere. It accommodates an estimated 400 billion stars, nebulae, and a wide range of interstellar gases and dust clouds. The merging light emissions of innumerable space entities are what furnish the Milky Way with its distinctive sheen.

When you observe our sprawling galaxy's contours in the night sky, you are literally affixing eyes on its defining feature, the stellar disk. The majority of the brightest stars in our galaxy, including the sun, are contained within this disk.

© Wikipedia Commons | Jeff Barton

Of course, it's impossible to wholly ascertain the disk's makeup because we are, in reality, on the inside looking out. Nevertheless, astronomers can say with absolute confidence that the disk is home to multiple billions of stars and copious amounts of dust. The considerable cosmic terrain between and among these untold stars is known as the "interstellar medium."

Summer evenings showcase the Milky Way imposingly climbing high in the night sky and spanning the constellations Cygnus and Aquila. To fully appreciate our parent galaxy and hone in on a vast and diverse array of celestial inhabitants basking in its luster, locate the outer-space region that hosts its radiant galactic center.

During the summer months, the zodiac constellations of Sagittarius and Scorpius, nicely visible on the southern horizon, are vibrant space neighborhoods chock-full of stargazing prey. The Milky Way's brilliant nucleus paints an ultrabright and impressive celestial portrait here—a warm and reassuring spectacle that is, ironically, a blisteringly hot and exceedingly violent sector of space.

In contrast with the summer snapshot of the Milky Way, the winter night sky serves up a conspicuously darker version of it. During wintertime, its disk is noticeably higher in the sky and

Winter: a prime and often overlooked stargazing season
© Thomas Nigro

positioned away from its dazzling center. Winter-night viewings of the Milky Way are still must-sees. The crisp, clear winter air, commingling with the dramatically darker galaxy higher overhead, furnishes stargazers with unmistakable visual contrast that cannot be gleaned in the summer months.

Stars

The countless stars visible in the night sky are literally flaming orbs of gases, mostly burning hydrogen and helium. The nearest star to Earth is, of course, the sun. It is scorching hot and in the midst of ongoing thermonuclear reactions. Substantial masses are what keep the sun and the family of stars from completely detonating and blowing apart during the furious nuclear reactions occurring in their interiors.

Our sun is what is known as a "second-generation star." At 4.6 billion years old it is not as old as the presumed age of the universe, and thus has a fair amount of living left to do. Scientists surmise that this yellow dwarf star has sufficient hydrogen to fuel its existence for several billions of years before it commences its celestial death rattle.

© Wikipedia Commons/NASA, Hubble European Space Agency, and Akira Fujii

Although our sun's diameter of 864,000 miles (1,390,000 kilometers) is immense, the biggest stars in the universe are one thousand times its size. The perceptible stars in the night sky are both sun-like and considerably larger. From an outer-space vantage point, the sun would appear as any other star—just one of an estimated 1 billion trillion in existence.

Although planetary bodies are sometimes confused with stars on the nighttime stage, they need not be. Planets merely reflect sunlight. Stars, like our sun, are veritable nuclear reactors in the extraterrestrial ether, converting hydrogen into helium throughout their enduring lives. They twinkle whereas planets do not. These pleasing ripple effects—refracted light—are courtesy of Earth's multiple layers of atmosphere, which twist faraway light sources like the proverbial pretzel.

The night-sky panorama provides you with a fine sampling of the diverse stars in the universe. Based on temperatures, the spectral classification of stars is blue-violet, blue-white, white,

© Ivo Laurin | Dreamstime.com

yellow-white, yellow, orange, and orange-red. The hottest stars are blue-violet; the coolest, orange-red. However, the high temperature of a star doesn't equate to luminosity.

The ultrabright Betelgeuse, for example, is a cool red giant star at the end of its celestial rope. Located in the constellation Orion the Hunter, Betelgeuse is one of the largest and most visible stars in the night sky. To find it, cast your eyes, binoculars, or telescope toward the mighty hunter's right shoulder. Estimates place its brightness capacity at 85,000 to 105,000 times that of our sun. Betelgeuse is 640 light-years away.

With new stars constantly forming and old stars petering out and disbanding as they exhaust the last of their energy sources, the universe is a work in progress. Although most stars appear to rise in the eastern sky and set in the western sky, some have a different modus operandi altogether. They are known as "circumpolar stars" and outwardly move in circles around a fixed point—like the celestial poles—whereas Polaris, the North Star, doesn't move at all.

The evident stars on the nighttime stage are also widely varying distances from Earth. Never lose sight of the fact that neighbors on the celestial sphere are visual neighbors. Although it may appear otherwise, they are rarely literal stellar neighbors.

For some distance perspective: Proxima Centauri, the nearest star outside of our solar system, is 4.2 light-years away from Earth. *Voyager 1,* which is the NASA space probe located as far away from Earth as any human-built contraption, is some 14 light-*hours* away. It has journeyed forty-plus years to reach what is believed to be the heliopause, the periphery of our solar system, and would require another 18,000 years of further exploration to be just 1 light-year away from Earth.

Deep-Sky Objects

A popular pursuit of many amateur astronomers involves hunting down what are classified as "deep-sky objects"—that is, celestial bodies beyond the sun, moon, and planets Mercury, Venus, Mars, Jupiter, Saturn, Uranus, and Neptune. Categories of deep-sky

Courtesy of NOAO/AURA/NSF/T.Rect or & B.A.Wolpa

objects include open clusters of stars, globular clusters of stars, nebulae, and galaxies.

Take the Andromeda Galaxy, M31, for instance, which is considered comparable in makeup to our Milky Way. It is centrally positioned in the constellation of the same name. At best, the naked-eye perspective of the Andromeda Galaxy is little more than an illuminated blur. Binoculars bring the image closer to you. However, telescopes are recommended for viewing our galactic neighbor and its very defined and imposing disk shape.

As a hobbyist interested in deciphering the night sky, you will encounter multiple space objects identified with the letter "M" followed by a number. The "M" stands for "Messier." Charles Messier (1730–1817) was a French astronomer who put deep-sky objects on the star map, if you will. He cataloged his sightings, which included star clusters, nebulae, and galaxies. Each one was assigned a number that is still in use today.

CHAPTER 4 Navigating Space Neighborhoods

Celestial Sphere

The celestial sphere is a snapshot of the night sky—not a literal one, but more of an imaginary friend to astronomers. It embodies a series of invented delineations, including a celestial equator slicing across its midsection, the north celestial pole at its apex, and the south celestial pole identifying its nadir. At less than 1 degree off the north celestial pole sits Polaris, the North Star.

In essence, the celestial sphere is a globular illustration that enables you to visualize the impact of Earth's rotation on objects in the night sky that appear and inevitably disappear. Most visible stars, for instance, mosey along from east to west across the

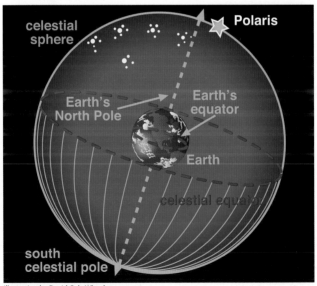

Illustration by David Cole Wheeler

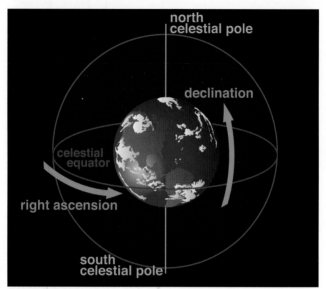

Illustration by David Cole Wheeler

celestial sphere on a daily basis. Their apparent motion takes them around the two celestial poles every night. This isn't because they are soaring through outer space at warp speed. Quite the contrary: Earth's unfaltering rotation on its axis is the wind beneath their wings.

To get a proper handle on the outer-space picture show, it is beneficial to envision celestial objects on one sphere based on their predictable locations at any given moment in time.

To assist them in decoding the celestial sphere's perpetual dynamism, astronomers utilize a coordinate system with right ascension (RA), the celestial-sphere equivalent of longitude on Earth's surface, and declination (DEC), the celestial-sphere equivalent of latitude on Earth's surface. Earth, remember, is always at the center of the celestial sphere, and we can spy no more than half of it from our surface vantage point.

Both right ascension and declination help to pinpoint where space bodies dwell. It enables professionals and amateurs alike to better understand, from moment to moment, what is transpiring in the night sky and where space objects are in relation to one another. The observable stars and other celestial bodies are widely varying distances from Earth but nonetheless appear as a uniform scene on the celestial sphere.

Our planet's rotation is west to east. This counterclockwise direction is how it appears from an outer-space catbird's seat peering down at the North Pole. However, we feel stationary on Earth's soothing terra firma, and thus envision sky movements in reverse. And because our planet completely rotates on its axis each day, the fundamental night-sky dynamics return for daily encores.

Constellations and Asterisms

The celestial sphere hosts numerous constellations of stars within its amphitheater. The International Astronomical Union (IAU) officially recognizes eighty-eight constellations in total. Forty-eight of them are considered the "classics," with boundaries established long ago by ancient Greek and Roman skywatchers. Fashioned by human beings to assist them in navigating the sprawling celestial ether, constellations are explicit regions of the night sky with generally accepted borders.

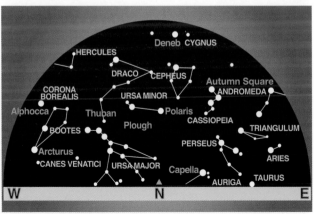

Illustration by David Cole Wheeler

No matter your location or the time of year, the night sky is chock-full of stars nestled in these celestial neighborhoods. Most of the constellations appear to rise in the east and set in the west—diurnal motion. As Earth simultaneously spins on its axis and orbits the sun, the evening picture window continually remakes itself as the days pass and the calendar turns. Constellations are further categorized seasonally based on their prime visibility to observers.

There are also constellations that do not ever "rise" or "set." These are the circumpolar constellations—Ursa Major, Ursa Minor, Draco, Cassiopeia, and Cepheus—that are visible throughout the year. Circumpolar constellations revolve around a fixed point in space.

There are certain times of the year when individual constellations are positioned in the night sky for very favorable viewings. The so-called winter constellations of the Northern Hemisphere include Auriga, Canis Major, Canis Minor, Cetus,

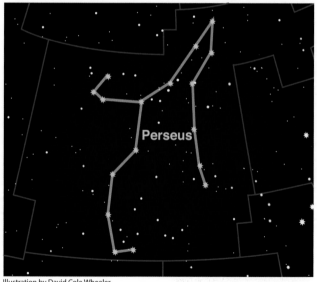

Illustration by David Cole Wheeler

Crater, Eridanus, Gemini, Lynx, Orion, Perseus, Sextans, and Taurus. If you can withstand the colder temperatures, it's highly recommended that you explore the winter night skies. The crisp, cleaner air of wintertime plays fewer hazing pranks on our planet's atmosphere, offering a clearer, less distorted portal into the celestial beyond.

The summer night skies in the Northern Hemisphere are likewise teeming with prominent and distinctive constellations to inspect. Stargazers are supplied with first-rate views of such summer constellations as Aquarius, Aquila, Capricornus, Cepheus, Corona Borealis, Cygnus, Delphinus, Lyra, Pegasus, Sagitta, Sagittarius, Scorpius, Scutum, and Vulpecula.

And one footnote here: On its annual journey known as an "ecliptic," the sun seemingly completes a full ring around the sky. It appears to pass through various constellations on this year-long odyssey. The constellations that host these ecliptic visitations are known as the "zodiac constellations." The sun, moon, and planets in our solar system are always on or nearby this well-lit path.

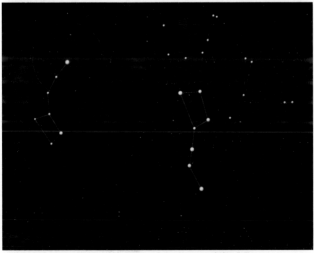

© Wikipedia Commons/NASA, Hubble European Space Agency, and Akira Fujii

The term "asterism" is applied to smaller patterns of stars that aren't constellations in their own right. In fact, asterisms are located within the confines of distinct constellations—like the Big Dipper in Ursa Major—or spread over one or more of them. Well-known members of the asterism fraternity include the Belt of Orion, the Northern Cross, the Pleiades, the Summer Triangle, and the Square of Pegasus.

Asterisms are highly useful in night-sky navigation. They serve as key markers on the nighttime stage, and can be readily spotted courtesy of their multiple-star configurations. For example, the aforementioned Ursa Major, the third-largest constellation, is one of the most effortlessly found because it is home to the celebrated asterism known as the Big Dipper.

Indeed, by first locating the Big Dipper with its unique bowl shape and two "pointer" stars, you are setting the stage for a further celestial adventure, as these stars pinpoint the position of Polaris, the North Star. Polaris can thereafter be your quintessential

Illustration by David Cole Wheeler

stargazing point of reference. The more you learn about these and other strategic celestial markers, the more rewarding your astronomical experiences will be.

Once upon a time human observers conceived constellations' celestial boundaries and assigned them names based on star patterns that reminded them of this, that, and the other thing. However, for most amateur astronomers, the figures at the foundations of constellations' names are ambiguous at best. With several notable exceptions, they are difficult, if not impossible, to decipher based on their alleged contours.

Camelopardalis the Giraffe, for instance, would hardly jump out at you as resembling a giraffe in the night sky. In order to discern the constellation's giraffe outline, you would first have to locate many faint stars within its borders. And even then, it might be straining credulity to envision a connect-the-dots giraffe among these stars. This is the case with most constellations.

Many of the eighty-eight constellations—notably the classic forty-eight—have fascinating and even Byzantine etymologies. For example, Lupus the Wolf, an obscure constellation in the southern sky, was not immediately identified as a wolf. It wasn't until the sixteenth century that it received its moniker. Prior to that time, Lupus was considered a generic wild animal. The neighboring Centaurus the Centaur is presumed to be carrying this slain animal in his arms.

Prominent Stars

Courtesy of the sun's proximity to Earth—some 93 million miles of separation—it appears considerable in size and extremely bright in the sky. It is the focal point of what we call a "solar system," a region in space under the sway of a particular star. With its substantial mass—almost 99 percent of our entire solar system—the sun's gravitational influence dominates all objects around it.

The sun's uncontested force completely choreographs the heliosphere, which reaches beyond Neptune's orbit to the

The sun: Earth's star of stars
© Thomas Nigro

celestial corner known as the heliopause. Here, the sun's solar wind wanes and another star's gravitas eventually takes over.

Recently, NASA's planet-hunting spacecraft Kepler uncovered a sun-like star with six planets orbiting it. This exciting find added to a growing number of known extraplanetary systems beyond our solar system. There are currently 3,090 confirmed systems and 4,160 confirmed "exoplanets," which are planets orbiting a star other than our sun in a unique solar system. With ever-increasing advances in technology, these numbers are expected to skyrocket in the coming years. So, when you cast your sights to the stars on high, appreciate the width and breadth of the celestial picture. The sun may be our lucky star, but it's just one among countless others, many of which have orbiting planets. And where there are planets, there may be life forms as well.

Skillfully navigating the night sky asks that you discern individual stars, series of stars (asterisms), and constellations of stars. Orion the Hunter, for example, is arguably the most recognized constellation in the night sky. Its unique star patterns,

Atmosphere of Betelgeuse · Alpha Orionis
Hubble Space Telescope · Faint Object Camera

Courtesy of Andrea Dupree (Harvard-Smithsonian CfA), Ronald Gilliland, NASA, ESA

which in fact resemble a hunter with a crossbow—a prominent exception in the constellation name game—are easily spotted and supply a welcome starting point from which to commence further celestial foraging.

Orion also accommodates an asterism known as the Belt of Orion—three very bright and very visible stars in a neat row—and is home to two of the brightest stars in the night sky: Rigel and Betelgeuse. Once you locate this celestial belt, the hunter in sum materializes before your eyes.

On the other hand, the Summer Triangle is an asterism containing three bright stars from three different constellations. It is seen high in the northern sky throughout the summer months. In fact, its three stars are all famous in their own right: Vega of Lyra, Deneb of Cygnus, and Altair of Aquila. From July through December, Vega—the brightest of the troika—is practically overhead during the twilight hours.

Star hopping is a technique seasoned stargazers employ to assist them in maneuvering to and fro in the immense night sky.

Courtesy of NASA and A. Fujii

It entails pinpointing the brightest and most familiar objects in outer space, knowing exactly what they are, and—yes—what is around them in all directions.

Take Aldebaran, the brightest star in Taurus the Bull, which is a fine wintertime sighting. To locate it, follow the Belt of Orion from right to left and continue on a vertical line. The shimmering Aldebaran is along this path. Distinctly orange in color, Aldebaran is sixty-five light-years away and forty times the size of our sun. Its distance explains why—considering the star's massiveness—it looks as small as it does. Aldebaran is part of an asterism in Taurus known as the Bull's Head. Appropriately, it is the bull's eye.

Star maps can also be extraordinarily helpful in getting to know stellar neighborhoods. In fact, once you know where things are—and where exactly to look for them—you will find that you waste much less time. On-demand star map Internet sites will print out charts based on your exact observational locations and specific dates—and you can't ask for more than that.

CHAPTER 5

Leading Constellations

Summer
Ursa Major the Great Bear

Illustration by David Cole Wheeler

Ursa Major the Great Bear is a circumpolar constellation, which means that it neither rises nor sets, as it were, and is discernible throughout the entire year from most Northern Hemisphere locations. The spring and summer seasons are nonetheless the best times to survey this constellation's sprawling terrain, which includes the Big Dipper among its celestial tenants.

Once you get a fix on the Big Dipper, follow along its starlit handle into the open sky. You will eventually encounter the exceptionally bright star Arcturus, which resides in the constellation Bootes. Further foraging in the same general direction leads you to the doorstep of another bright star, Spica, in the zodiac constellation Virgo. There is in fact a maxim to help you remember this stargazing lateral: "Arc to Arcturus and Speed to Spica."

While in the neighborhood, also explore the stretch of sky below the two stars in the Big Dipper's cup. Regulus, the brightest star in the zodiac constellation Leo, awaits you. The far-reaching Ursa Major borders on multiple constellations: Draco, Camelopardalis, Lynx, Leo Minor, Leo, Coma Berenices, Canes Venatici, and Bootes.

Corona Borealis, the Northern Crown

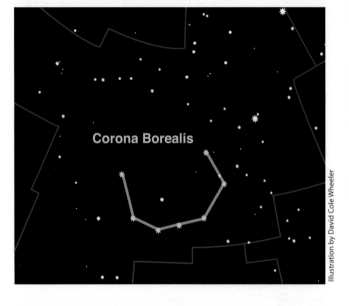

Illustration by David Cole Wheeler

Although Corona Borealis, the Northern Crown, is a relatively small constellation in the northern sky, it is nonetheless a fine spring and summertime attraction. This is because it accommodates an alluring and distinctive asterism in the shape of a neat semicircle.

Alphecca is concurrently the brightest star and the centerpiece of what is popularly known as the Northern Crown. Smack dab in the middle of this exceptional half-circle of stars, it distinguishes itself as the celestial crown's shimmering white jewel.

Situated between the constellations of Bootes and Hercules—and two of the night sky's brightest stars, Arcturus and Vega—the Northern Crown is easily located. As stand-alone stars, Arcturus, with its orange-reddish hues and luster, and Vega, which sparkles a hospitable bluish-white, are stargazing must-sees.

Corona Borealis also houses the curious variable star, R Coronae Borealis, which occasionally disappears from view when carbon condensation swaddles it in stellar grunge. A variable star is so named because of its varying degrees of brightness as detected from Earth. These variations can be the consequence of numerous factors, including changes in a star's illumination due to internal disruptions or completely extraneous factors in its surrounding space neighborhood.

Draco the Dragon

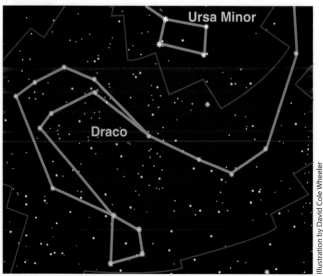

Illustration by David Cole Wheeler

Revolving around the North Pole as it does, Draco the Dragon is only observable in the Northern Hemisphere. Draco is a circumpolar constellation and thus visible throughout the entire year. However, the summer and fall seasons are the best times to check out this area of sky rich in mythological underpinnings.

Draco's elongated and twisting borders converge with multiple neighboring constellations: Bootes, Hercules, Lyra, Cygnus, Ursa Minor, Camelopardalis, and Ursa Major. To locate Draco in the night sky, hunt for the dragon's head, if you will, which is a trapezoid of stars found north of the constellation Hercules. Meanwhile, the celestial dragon's long and winding tail of stars meanders far and wide, finishing up somewhere between the Big Dipper and Little Dipper, which are in the constellations Ursa Major and Ursa Minor respectively.

An interesting find on the constellation's periphery is the star Thuban, the former North Star. Draco also contains the Cat's Eye Nebula, NGC 6543, on its celestial turf. "NGC," by the way, is the acronym for "New General Catalogue," another widely used cataloging system of space bodies.

The Summer Triangle

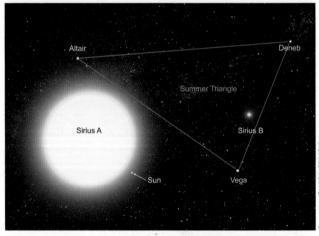

Courtesy of NASA/ESA/G. Bacon

As you can glean from its title, the Summer Triangle is prominent in the night sky during the dog days of July and August. It is not a constellation but an asterism made up of three eminent stars—three stars, by the way, in three different summer constellations: Altair from Aquila, Deneb from Cygnus, and Vega from Lyra.

Indeed, the Summer Triangle of stars graces the southeast horizon during summertime. Its considerable size explains why it so effortlessly spans three constellations. The aforementioned stars are nonetheless all very bright and thus easy to pinpoint in the ether. When poring over this region of the summer night sky, it's impossible to miss these three bright stars that form three points of a stellar triangle.

As some added stargazing windfall at this time of year, the Milky Way dramatically wends its way behind the Summer Triangle. Beginning in Cassiopeia in the northeast sky, our home galaxy cuts an extended, noticeably bright but nonetheless fuzzy swath directly across the triangle to Scorpius in the southwest. In the neighborhoods of Scorpius and Sagittarius the galaxy's spirited nucleus struts its hot stuff.

Cassiopeia the Queen

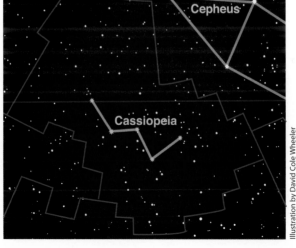

Illustration by David Cole Wheeler

Cassiopeia the Queen is a well-known northern sky constellation. It is also among the earliest recognized and named constellations. The ancients were keenly aware of this snippet of night sky. Indeed, during optimal viewing times in the summer and fall, the circumpolar Cassiopeia, courtesy of its five bright stars forming an unmistakable "W," is readily spied on the celestial sphere.

To locate Cassiopeia, zero in on Polaris, the North Star, and begin tracing an imaginary line due south. This course will initially pass through the constellation Cepheus before encountering a group of stars in a "W" pattern. Continuing south on this same celestial path puts you in the vicinity of the Andromeda Galaxy. There are also two Messier objects worth checking out in Cassiopeia: M52 and M103. Both of these globular clusters of stars can be nicely seen with binoculars or, better still, with telescopes.

Cassiopeia is one of many constellations named for figures out of Greek mythology. Her daughter was Andromeda, which not coincidentally borders Cassiopeia to the south, while Cepheus reigns to Cassiopeia's north. Other bordering constellations are Camelopardalis, Lacerta, and Perseus.

Cygnus the Swan

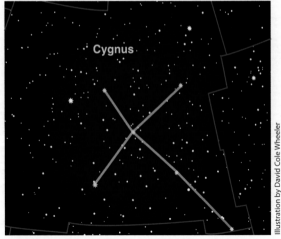

Illustration by David Cole Wheeler

Cygnus the Swan is a northern constellation visible high in the summertime night sky. It is best known for accommodating the Northern Cross. The Northern Cross is an asterism of stars and should not be confused with the more famous Southern Cross in the Southern Hemisphere. The bordering constellations of Cygnus, which very fortuitously resides in the Milky Way's warm embrace, are Cepheus, Draco, Lyra, Vulpecula, Pegasus, and Lacerta.

The Northern Cross boasts an intriguing double star atop it known as Albireo. With the aid of binoculars or a telescope—even a small one—Albireo's double life is unmasked. Whereas the naked eye perceives just one, you can clearly distinguish two stars marking the cross's high point, with one orange in color and the other blue.

The equally bright anchor of the Northern Cross is Deneb, a lone star. Deneb, by the way, performs double duty as a star in another familiar asterism and favorite of summertime stargazers, the Summer Triangle. Deneb is believed to be twenty-five times the sun's dimensions and to shine 60,000 times brighter.

Winter
Orion the Hunter

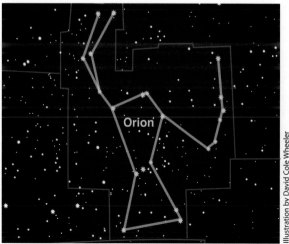

Illustration by David Cole Wheeler

The prominent winter constellation Orion the Hunter is a personal favorite of many amateur astronomers. To get a fix on this outer-space terrain, cast your eyes eastward and locate a quadrilateral containing three stars in a neat row. This is the Belt of Orion, which is positioned more or less in the center of a very busy and very distinctive constellation. On opposite ends of the belt are two very bright stars as seen from our earthly vantage point: Betelgeuse and Rigel.

The celestial finds in Orion don't begin and end with the mystical hunter's star-studded belt. Immediately below the belt of stars is the Orion Nebula, M42, which exudes a well-dispersed and conspicuous glow in the crisp and clear winter night sky. In the celestial shadows of this cavernous, dusty, and gaseous region of space is a fertile breeding ground for multiple thousands of stars.

In fact, massive young stars fashion the colors and contours of the Orion Nebula, which are nicely discernible in wintertime. This nebula is both the nearest—at 1,500 light-years away from Earth—and brightest star-breeding ground in view.

Canis Major the Big Dog

Illustration by David Cole Wheeler

Canis Major the Big Dog is located south and east of the prominent constellation Orion. Look for Canis Major low in the winter sky, practically hugging the southern horizon. Greek mythology has Canis Major as one of the great hunter Orion's canine companions. This explains why the constellation is literally nipping at the heels of Orion in the night sky, although it doesn't physically border its celestial neighbor.

Situated along the wintertime Milky Way, Canis Major accommodates multiple bright stars on its stellar turf, including Sirius, the brightest star in the Northern Hemisphere. If you look just below the mighty hunter's belt, you will be in the environs of Canis Major. The radiant Sirius will alert you in no uncertain terms that you've found what you are after.

The legendary Sirius, nicknamed the "Dog Star," is situated where the pooch's collar would approximately lie in a connect-the-stars canine. Sirius is also part of the asterism called the Winter Triangle. Canis Major shares a border with constellations Monoceros, Lepus, Columbia, and Puppis. Appreciably to the north of Canis Major is Canis Minor, known as the "Lesser Dog."

Andromeda the Chained Maiden

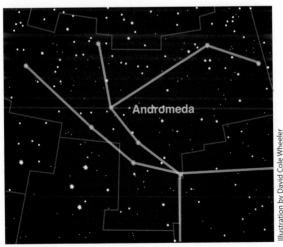

Illustration by David Cole Wheeler

Although it is visible in the summer months, Andromeda the Chained Maiden is ideally positioned in the Northern Hemisphere's fall and winter. The constellation is curiously V-shaped and hovers precariously close to the north celestial pole.

Andromeda's most renowned tenant is our galactic neighbor, the Andromeda Galaxy, M31, a sprawling spiral galaxy much like our Milky Way in composition, albeit larger. The galaxy is centrally located in Andromeda and can be seen—under highly favorable conditions—with the naked eye. However, a telescope is required to appreciate the galaxy's shapely celestial form.

While you check out Andromeda, it's also a propitious time to visit the Great Square of Pegasus, which, outside of the Big Dipper, is probably the most recognizable asterism of stars in the Northern Hemisphere. The four stars that form this stellar square are actually in two separate constellations: three in Pegasus and one in Andromeda. The stars are Scheat—a reddish giant at its apex—and Markab, Algenib, and Alpheratz. Alpheratz calls Andromeda home. Because its interior appears largely starless, the Great Square of Pegasus strikes a singular celestial pose. In addition to Pegasus, the constellations that border Andromeda are Perseus, Cassiopeia, Lacerta, Pisces, and Triangulum.

Cetus the Whale

Cetus the Whale is a fall and winter constellation in an area of the sky associated with water and water-related things. It is among the biggest constellations—only three cover more celestial ground than Cetus. Although it is not a zodiac constellation, the sun's ecliptic passes very near its borders. Thus, planetary visibility within Cetus occurs for short spells.

Cetus also has the distinction of hosting Mira, the first variable star ever discovered. Mira at times is among the brightest stars in the nighttime sky, and on other occasions barely visible. The discovery of this star in the seventeenth century greatly contributed to the advancement of astronomy. Prior to detecting Mira's mercurial behavior, science viewed the celestial heavens as unchanging—invariable, as it were.

Cetus shares borders with Aries, Pisces, Aquarius, Sculptor, Fornax, Eridanus, and Taurus. It also is situated a great distance from the Milky Way's plane and the obscuring dust therein, which cloaks innumerable deep-sky objects. Thus offering a very clear portal into the night sky, there are multiple faraway galaxies visible within Cetus.

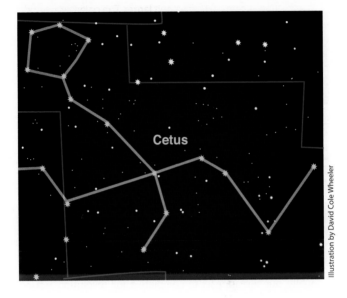

Illustration by David Cole Wheeler

CHAPTER 6 Zodiac Constellations

Summer
Leo the Lion

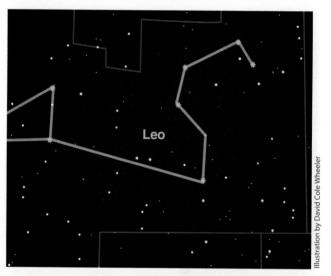

Illustration by David Cole Wheeler

Annually, our sun appears to complete a full revolution around the sky on an unremitting eastward track known as the ecliptic. The regions of sky on this trajectory are called the zodiac constellations. In fact, Leo the Lion was one of the earliest identified and drawn constellations. The stars in this, the fifth zodiac constellation, with a little imagination, resemble a crouching lion gazing westward. Leo is a sprawling constellation with multiple neighbors touching its space borders: Ursa Major, Leo Minor, Lynx, Cancer, Hydra, Sextans, Crater, Virgo, and Coma Berenices.

Leo is prominent in the spring and summer night skies and accommodates many bright stars, including Regulus, which can be spotted south of the Big Dipper's pointer stars. Look, too, for stars Denebola and Leonis.

Leo counts multiple galaxies on its home turf, including M65, M66, and NGC 3628, which comprise what is known as the Leo Triplet. This grouping can be observed with most types of telescopes. There is also the Leo Ring to check out: a primordial cloud of gases—hydrogen and helium—believed to be a remnant of the universe's explosive and epic beginnings.

Virgo the Maiden

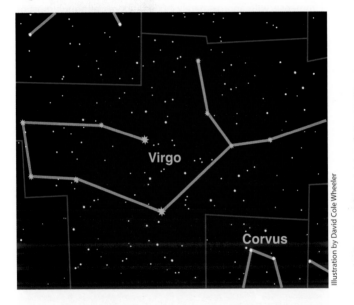

Illustration by David Cole Wheeler

Virgo the Maiden is the second-largest constellation in the night sky and one of the more happening zodiac constellations. Spring and summer are the optimum times to explore this teeming celestial expanse. Virgo is located to the west of Leo, with Libra to its east. Spica, its brightest star, makes finding the constellation's celestial address relatively simple. The arc of the Big Dipper leads the way, first to Arcturus in the constellation Bootes, and then to Spica in Virgo.

Virgo's boundaries also mark one of the points where the celestial equator intersects with the ecliptic. The sun actually remains in Virgo country for forty-five days. This sprawling constellation shares borders with numerous neighbors: Bootes, Coma Berenices, Leo, Crater, Corvus, Hydra, Libra, and Serpens Caput.

Eleven Messier objects call Virgo home. Most of them are galaxies in what is known as the Virgo Cluster. This intergalactic neighborhood, which crosses over into the constellation Coma Berenices, is immense, with approximately 2,000 residents, although most of them are considered "dwarf galaxies." Virgo also has the unique distinction of hosting more known extrasolar planets—planets orbiting stars other than the sun—than any other constellation.

Scorpius the Scorpion

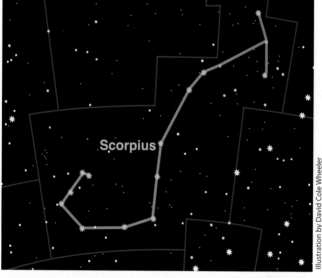

Illustration by David Cole Wheeler

Scorpius the Scorpion is at once a very large and easily found constellation. Neighbors include Sagittarius, Ophiuchus, and Libra. For amateur astronomers, Scorpius is a prominent patch of night sky worth exploring in detail.

The constellation hugs the horizon in the Northern Hemisphere's summertime. Among the exceptions to the rule in the family of constellations, Scorpius's distinctive star formations quite literally resemble a scorpion. Indeed, its brightest stars resemble something of a celestial arachnid. Antares, a red supergiant star in Scorpius, is easily discernible in the belly of the beast.

Scorpius is also an area of space in close proximity to our parent galaxy's galactic center. Based on its fortuitous coordinates in the night sky, its inhabitants greatly benefit from the Milky Way's illuminated embrace. Scorpius's well-lit celestial terrain supplies you with a fine view of many deep-sky objects, which in darker regions of space would go unseen. One of the closest to Earth and most impressive clusters of stars, globular cluster M4 is found in here. Other deep-sky objects to look for in Scorpius include the Butterfly Cluster, M6, and the Ptolemy Cluster, M7.

Sagittarius the Archer

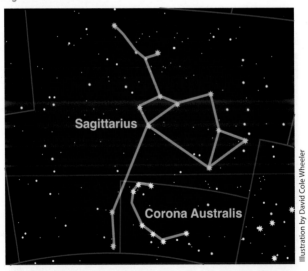

Illustration by David Cole Wheeler

With the ecliptic ushering the sun through Sagittarius the Archer in early wintertime, this active constellation is best observed low in the sky during the summer months. It is a region of space teeming with highly visible celestial objects. For one, it is on the ecliptic and thus regularly plays host to the moon and myriad planets.

As viewed from our earthly perch, Sagittarius is also home to the very brightest stretch of the Milky Way, including its galactic center at some 27,000 light-years away. Accommodating the nucleus within its boundaries ensures that this zodiac constellation is a lively expanse of the summertime night sky. Sagittarius's fertile, creamy-looking star fields are the by-product of this exclusive celestial location in the densest area of our home galaxy.

Indeed, the galaxy's tender glow enables deep-sky objects such as nebulae and star clusters to reveal themselves in considerable numbers here. There are fifteen Messier objects in Sagittarius. The bright and bulky globular cluster known as M55 is a favorite stargazing target, as is the Lagoon Nebula, M8. There are more individual stars with known planets orbiting them in Sagittarius than in any other constellation.

Winter
Taurus the Bull

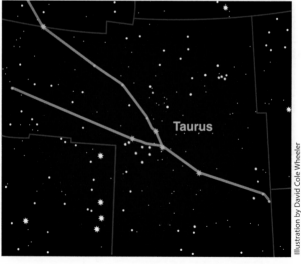

Illustration by David Cole Wheeler

Taurus the Bull, which appears to be charging the great hunter Orion on the celestial sphere in wintertime, is a zodiac constellation of great renown. Lying just southwest of Orion, Taurus is one of the more recognizable regions of space at this time of year with its well-defined V-shape of stars. In addition to Orion, Taurus shares borders with several other constellations: Auriga, Perseus, Aries, Cetus, Eridanus, and Gemini.

The easily spotted, very radiant orange star, Aldebaran, lies near the ecliptic and serves as an admirable celestial bull's eye. Taurus's horns and head are part and parcel of the Hyades open cluster of stars. The cluster's brightest stars, coupled with Aldebaran, form that conspicuous V in the night sky. Although it appears otherwise, Aldebaran, which is much closer to Earth, is not part of the more distant cluster.

Also in Taurus is the famous Pleiades Cluster, M45, aka the "Seven Sisters" of Greek mythology, which is a dazzling open cluster of stars. Look for it northwest of the radiant Aldebaran. Although this cluster embodies hundreds of stars, only fourteen can be seen with the naked eye, and even then, only under optimum stargazing conditions.

Gemini the Twins

Following on the heels of Aries and Taurus, the sun's ecliptic ushers it into the zodiac constellation of Gemini the Twins. Just due west of Taurus, Gemini is prominent in the winter night sky, with Cancer the Crab to its east. It can be easily located with the unaided eye.

Almost directly overhead in midwinter, its celestial terrain is observable to the mighty hunter Orion's left. The constellations of Auriga and Lynx reside to Gemini's north, with Monoceros and Canis Minor rounding out its southern flank.

Of course, Gemini is renowned for hosting the twin stars of Castor and Pollux, which appear very near each other on the celestial sphere, and are at the foundation of the constellation's legendary nickname. While these two bright stars look like close neighbors in the sprawling night sky, they are in

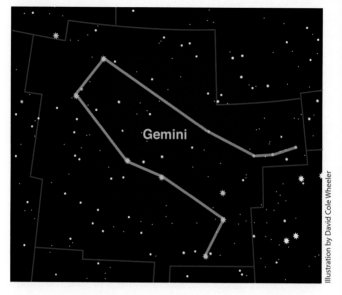

Illustration by David Cole Wheeler

fact very far apart. Nevertheless, it's their visible closeness that inspired the mythology of the twins. And when you factor into the stellar equation some of the less-bright stars sloping downward from Castor and Pollux, one could conceivably imagine a connect-the-stars resemblance to two human stick figures.

CHAPTER 7

Star Search

Closest Stars

Excluding the sun, the nearest star to Earth is Proxima Centauri, aka Alpha Centauri C. Classified as a red dwarf star, it is part of a triple star system and approximately 4.2 light-years away. Its companion stars Alpha Centauri A and B are 4.3 light-years away. Other stellar neighbors include Barnard's Star at 5.9 light-years away; Wolf 359 at 7.7 light-years away; Lalande 21185 at 8.3 light-years away; Sirius A and B at 8.6 light-years away; and A and B Luyten 726-8 at 8.7 light-years away.

From our earthly vantage point, the nearest stars to our planet and solar system aren't always the brightest stars in the

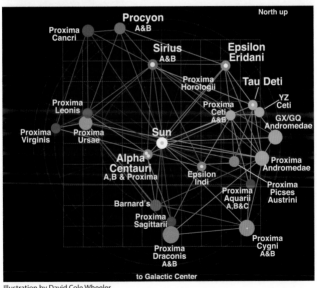

Illustration by David Cole Wheeler

Courtesy of NASA/CXC/SAO

night sky. A star's distance from Earth is merely one component in a multilayered celestial equation.

Indeed, the fraternity of stars in the universe is by no means monolithic in size, composition, and—yes—illumination. Many of the stars in close proximity to our solar system are invisible to the naked eye and observable only with a pair of binoculars or a telescope. These decidedly muted stars are typically red dwarfs, which are often too dim to see, from any distance, without an enhancing device.

Proxima Centauri, the nearest star outside of our solar system, is a rather pedestrian outer-space body. But the fact that it's the closest star to Earth makes it popular stargazing prey. Just like celestial neighbors Barnard's Star and Wolf 359, Proxima Centauri is a red dwarf star and practically imperceptible in the night sky.

Proxima Centauri is part of the Alpha Centauri system of stars. It orbits two companions, Alpha Centauri A and Alpha Centauri B, and can be found in the constellation Centaurus. Meanwhile,

Barnard's Star can be tracked down in Ophiuchus the Serpent Holder, but is not visible to the unaided eye. You will need binoculars or a telescope to locate this simultaneously very near and very faint star.

Red dwarf stars are the most common stars found in our stellar neck of the woods. In unmistakable contrast with other types of stars, red dwarfs have very low surface temperatures. They nonetheless endure for a very long time, which accounts for their ubiquity. It is believed that red dwarf stars can soldier on for trillions of years—more than the presumed age of the universe.

Brightest Stars

From our perch here on Earth's surface, the sun is the brightest star around, with no close second. At a mere 93 million miles away, we can practically reach out and touch it. Many of the most luminescent stars in the night sky are quite a distance away from our planet. For example, Rigel, among the top ten brightest stars from

Stargazing bookends: sunset and sunrise
© Thomas Nigro

our vantage point, is 775 light-years away. And while nearness to our planet doesn't equate to brightness, distance plays a critical role in the overall equation.

Many stars are exceptionally vivid but don't appear so because they are very far away. Others are conspicuously less vibrant but relatively close to us, and therefore distinguish themselves on the celestial sphere.

With the sun measuring 1 in illumination—the standard that all other stars are measured against—Sirius scores an impressive 23, which means that it shines 23 times brighter than the sun. At 8.6 light-years away, Sirius is highly visible in the constellation Canis Major. The runner-ups to Sirius are Canopus in Carina; Alpha Centauri A and B, aka Rigil Kentaurus, at the foot of the constellation Centaurus; Arcturus, the brightest star in Bootes; and Vega in Lyra.

Summer is an ideal time of year to locate many of the brightest and most celebrated stars. To the east, you'll find Sirius in Canis

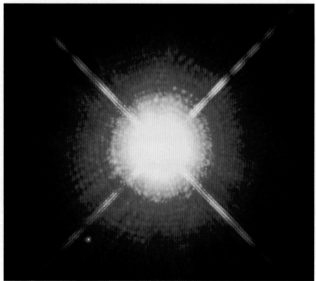

Courtesy of NASA [http://www.nasa.gov/], H.E. Bond and E. Nelan (Space Telescope Institute [http://www.stsci.edu] Baltimore, Md.); M. Barstow and M. Burleigh (University of Leicester, U.K.); and J.B. Holberg (University of Arizona)

Major. The Dog Star is close to our solar system as these things go, which naturally enhances its luminosity from our perspective. Under idyllic conditions, Sirius can be seen even in the daylight hours when the sun is low on the horizon.

Sirius is nearly twice as bright as Canopus, which receives the night sky's silver medal. However, Canopus is a tricky stargazing find due to its low sky positioning in the southern constellation Carina. Contrarily, Castor and Pollux, Gemini's "heavenly twins," and the constellation's two brightest stars, are worthwhile summer quarry. You'll discover, too, that Castor is somewhat fainter than Pollux.

Winter is likewise a hospitable season to hunt down bright stars, including the bluish-white Rigel, a supergiant in Orion. In early wintertime, Rigel rises in the east just after sunset. Another interesting seasonal sighting is the yellow-hued Capella, positioned midway in the northern sky's Auriga constellation, and part of a binary star system (i.e., multiple stars orbiting one another).

Star Clusters
Pleiades Cluster

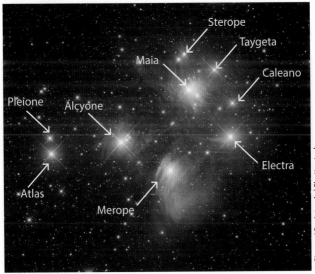

© Giovanni Benintende | Shutterstock

Star clusters, sometimes called "star clouds," are popular stargazing targets. In the annals of amateur astronomy, the Pleiades Cluster, M45, is the best known of open star clusters and has been since antiquity. The Pleiades Cluster is found in the constellation Taurus and is best observed during the winter months. To locate this impressive star assemblage, look northwest of the very bright Aldebaran, the orange bull's eye in Taurus. Although the cluster accommodates hundreds of stars, only fourteen are ever visible to the naked eye.

The cluster is 425 light-years from Earth, which explains why some of the extraordinarily bright stars amid the grouping appear rather faint from where we sit, and why so many of them cannot be seen without a telescope.

With its famous nickname of the "Seven Sisters," the Pleiades showcase seven corresponding and very bright stars—Sterope, Merope, Electra, Maia, Taygeta, Caleano, and Alcyone—plus two more: Pleione and Atlas, the mythological siblings' mother and father. Alcyone is estimated to shine a thousand times brighter than our sun! Blue reflection nebulae sheath many of these stars in an ethereal glow, which makes the Pleiades a must-see.

Double Cluster in Perseus

© peresanz | Shutterstock

The Double Cluster in the northern constellation of Perseus is comprised of two distinct open clusters of stars (NGC 869 and NGC 884). What makes the Double Cluster so unique is not only its dual nature, but the fact that the clusters are separated by only 300 to 400 light-years, the celestial equivalent of walking distance.

The Double Cluster can be observed with the naked eye in dark sky areas. Nevertheless, this magnificent celestial twofer is best viewed with a pair of binoculars or a telescope, allowing you a much sharper look at the hundreds of bright stars in the twin clusters.

The Double Cluster is a rare, prominent deep-sky object overlooked by French astronomer Charles Messier. This outer-space duo is circumpolar, never rising or setting, as it were. The clusters are visible above the horizon every night of every day of the year. However, the best time to explore the Double Cluster is in the fall and winter when it is high in the sky. To locate this prized deep-sky jewel—or jewels, in this case—look to the northernmost region of Perseus in close proximity to its border with Cassiopeia.

Great Globular Cluster

Courtesy of NASA

The Great Globular Cluster in Hercules, M13, is among the most sought-out deep-sky objects in the summer night sky. Under dark and otherwise ideal weather conditions, the cluster can be seen with the naked eye. However, binoculars and telescopes are recommended for viewing it.

The Great Globular Cluster is 20,000 light-years from Earth and around 145 light-years in diameter. Globular clusters typically consist of spherical congregations of aging stars with, if you will, a common ancestry. Nonetheless, there are some children in Great Globular Cluster's celestial house, which are the distinctive blue and white stars in the mix.

The hundreds of thousands of stars in the Great Globular Cluster are bound together by the Milky Way's awesome gravitational pluck. This is what sets it apart from open star clusters, which are considerably less at gravity's beck and call and more loosely dispersed in outer space. Edmond Halley first documented its existence in 1714, writing, "It shows itself to the naked eye when the sky is serene and the moon is absent"—sound advice then and now for observing the Great Globular Cluster and countless other celestial objects, both near and far.

CHAPTER 8 Moon Observation

Phases

If you would prefer to explore the cosmic ether a little closer to home, rather than outer-space bodies light-years away, the moon is there for the picking. Its utter dependability makes moon watching a whole lot more straightforward than hunting down faraway deep-sky objects.

The moon's recurring phases are the consequence of its ever-shifting angles as it revolves around our planet. Its position vis-à-vis both Earth and the sun is what illuminates, and sometimes obscures, parts or all of it. The moon's initial phase—called the "new moon"—occurs when the sun, moon, and Earth are in relatively close alignment, with the moon sandwiched in between and revealing only its dark, non-illuminated side to us. Except

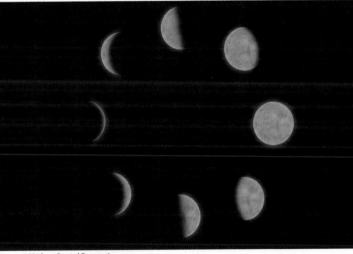

© Marleen Smets | Dreamstime.com

© Rgaf72 | Dreamstime.com

during rare daytime solar eclipses, the moon is completely imperceptible to our prying eyes during this opening salvo.

As the moon continues its orbit around Earth, it finds itself aligned once again—in approximately two weeks—with the sun and Earth. But Earth, and not the moon, is in the middle of this go-around. When the moon is positioned on the far side of our planet, we witness a full moon.

The moon is full of surprises. When you gaze into the early night or early morning skies during its crescent phases, you just might spy something curious. You may see the moon's normally vibrant crescent coupled with a view—in contrastingly diffused light—of its full orb. This phenomenon is known as "earthshine."

During a waxing crescent moon, earthshine is visible after sunset. During a waning crescent moon, earthshine occurs in the early morning hours. This celestial spectacle fashions an atmospheric—gently mysterious—aura in the skies. What, in fact, you are witnessing at these moments is the sun's light reflecting off our planet's surface to the moon's surface, which then casts the light back at us.

Earthshine occurs in close proximity to the imperceptible new moon phases because, at these times, Earth is fully illuminated from the perspective of a lunar horizon. In other words, if you were residing on the moon and checking out Earth in the nighttime sky, you would be seeing our planet's equivalent of a nearly full moon. This intense reflective sunlight from Earth to the moon back to Earth again supplies us with a celestial snapshot worth checking out.

Favorable Observation Times

Contrary to popular opinion, the best time to observe the moon is not during its full phase. Although it is a pleasing visual in the night sky, the full moon is extremely bright. Looking through binoculars or telescopes at a full moon is a blinding experience, downright painful to the eyes; its sheer brightness overwhelms all else.

Indeed, the intense sheen of the full moon makes surveying the distinctive and intricate features of our planet's only natural satellite well nigh impossible. The moon's concentrated luminescence in this phase gives it a flat look on the nighttime stage, almost like a one-dimensional Post-it note.

Dark and light: lunar maria and lunar highlands on full display
© Thomas Nigro

The preferred observational times are phases on either side of the new moon—just before or just after it. The crescent moon phase, for instance, embodies a visual moon at less than 50 percent illumination. Because of its unique curvature, this particular moon phase always garners attention from the skywatchers. The absence of excessive glare during crescent phases also encourages various night-sky couplings throughout the year, such as the moon alongside planet Mars and alongside planet Venus.

If you want to explore the minutiae of the lunar landscape, schedule your astronomical outings when there's a noticeable "terminator" slicing across it. The terminator is the line that separates night and day on the moon, and is most prominent on or near its quarter phases. There is no evident terminator during a full moon.

Approximately 17 percent of the moon's sprawling surface is covered with shadowy balsitic plains called the lunar maria. In contrast to the dark, relatively flat terrain that defines the lunar maria, the remainder of the moon's terrain—which happens to be

Half moons, not full moons: ideal times to explore Earth's sole natural satellite's intricate surface
© Thomas Nigro

its majority, at nearly 83 percent—consists of lunar highlands, or terrae. With a terminator in view, both the lunar maria and lunar highlands are cast into marked relief.

The waxing gibbous phase, for example, is a fine time to explore the moon's surface. During this phase, you can clearly decipher the terminator. The prior moon phases, waxing crescent and first quarter, are also favorable observation times. Keep in mind, too, that although it's called a "half moon," the first quarter moon is significantly less than half as bright as a full moon.

Lunar Maria

When ancient astronomers first explored the moon in the night sky, they observed flat, shadowy areas that contrasted with a more rugged—and more abundant—lighter surface. They assumed these darker hues were oceans and named them *maria,* or the singular *mare*—the Latin term for "seas." And although these sky-watchers from yesteryear were off beam in their guesswork, the moniker stuck.

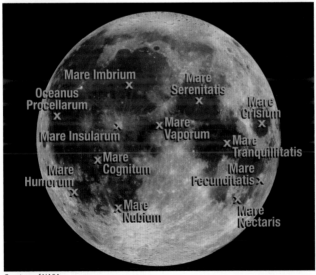

Courtesy of NASA

The lunar maria are volcanic plains believed to be the consequence of volcanic eruptions roughly 3 to 4.2 billion years ago. *Apollo 11* disembarked on a small lunar mare dubbed the Mare Tranquillitatis, aka the "Sea of Tranquility." In contrast to other maria, the Mare Tranquillitatis flaunts a lightly bluish tint, which is very likely the by-product of higher metal content in its rocky terrain. With their iron-intensive compositions, lunar maria are also poor reflectors of sunlight—hence their dark appearance. Lunar maria rocks are iron-rich basaltic lavas not unlike Hawaiian volcanic debris.

The various lunar maria throughout the moon's surface have been assigned names to differentiate them from one another. Names include Mare Imbrium (Sea of Rains), Mare Nectaris (Sea of Nectar), and Mare Nubium (Sea of Clouds).

Lunar Highlands

In sharp contrast to the lunar maria, the lunar highlands are light-colored and mountainous in appearance. The highlands are made

Courtesy of NASA

up of anorthosite, a rock consisting of lighter elements like calcium and aluminum, which explains the two-tone moon we see in the night sky. In fact, there are numerous highlands situated along the periphery of extensive areas of maria.

The scientific consensus is that the highlands originated during the moon's formation when feldspar crystallized and worked its way to the pinnacle of the oceans of molten lava generated by volcanic activity. Although both the maria and highlands are cratered, the latter are considerably more wracked with the residue of outer-space wreckage. Without Earth's protective layers of atmosphere, the moon has long been the recipient of a sustained barrage of meteoroid and asteroid strikes.

The moon's southern highlands are especially worth exploring, for they reveal a multifaceted surface with numerous overlapping craters. Perhaps more than any other area of the moon, this particular region—with its densely cratered mien—furnishes you with a stunning picture window into its spectacular desolation. Binoculars or a telescopic survey are recommended for this inside peep.

Craters

The moon and its craters go hand in glove. Indeed, this natural satellite of Earth is pockmarked by millions of craters of every conceivable size—from the minuscule found in individual rock pieces to some that span over 200 miles (360 kilometers) in diameter. There are over forty so-called "impact basins" on the moon's surface. Based on having rim diameters of 185 miles (300 kilometers) or more, craters receive not only this ballyhooed distinction but a name as well.

With a telescope, you can feast your eyes on the seemingly infinite series of craters that characterize the moon's surface. Check out the craters' distinctive features, such as raised rims, which indicate surface materials once upon a time expelled by a nearby impact. Also, explore their floors, relatively level areas that amass mini craters with the passage of time.

One intriguing crater to put on your stargazing must-see list is the Goclenius crater, which is situated near the outskirts of Mare

Courtesy of NASA

Fecunditatis. Its crater edges are conspicuously worn and asymmetrical in appearance. The floor of Goclenius also exhibits a visibly narrow, twisted, and protracted channel, which is indicative of past lava flows.

There is no wind or free-flowing water on the moon to erode its many craters. In other words, they remain largely as they were upon initial impact—that is, until another strike comes along and reconfigures the landscape.

The Copernicus lunar crater is one of the most frequently observed by amateur astronomers. It is located due northwest of the moon's center. This impressive crater is readily spotted with a pair of binoculars or, better yet, a small telescope. It is believed to be only 800 million years old—a mere youth as these things go. And because of its relatively young age, Copernicus—in contrast with older craters—reveals no evidence of having been flooded with molten lava.

Courtesy of NASA

When zeroing in on the moon's craters, take note that its surface sports a distinctive topsoil known as the regolith. With no protective atmosphere to fend off outer-space interlopers, ground-up rock fashions this singular celestial strain of dirt. During their various lunar missions, *Apollo* astronauts left visible footprints while moon-walking, revealing the presence of this loose, albeit rocky, soil atop the moon's desolate and cratered landscape.

CHAPTER 9 Planet Observation

The Planets

Our solar system accommodates eight officially designated planets, five dwarf planets (Pluto, Ceres, Eris, Haumea, and Makemake), and 200 or more natural satellites. While the sun sports a mean radius of 432,000 miles (696,000 kilometers), no planet or other solar system object even comes close. Jupiter, the radius runner-up, spans 43,440 miles (69,911 kilometers)—a puny gas ball relative to the sun. But size matters an awful lot in space, with smaller masses orbiting bigger ones.

On the planetary stage, Jupiter is nonetheless the king. While Earth is the largest of the terrestrial planets—those nearest the sun—it is positively minuscule when compared with Jupiter. It is estimated that 1,000 Earths could fit into Jupiter, and that

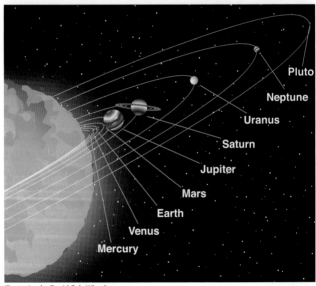

Pluto
Neptune
Uranus
Saturn
Jupiter
Mars
Earth
Venus
Mercury

Illustration by David Cole Wheeler

1,000 Jupiters could fit inside the sun. Yes, the same sun that is an average star vis-à-vis body mass in its considerable celestial fraternity.

Like the much more distant stars, planets seemingly meander across the night sky in an east-to-west direction, although sometimes their movements appear otherwise. Nevertheless, because the eight planets orbit the sun, they never stray far from its path—the ecliptic.

Stars often appear as twinkling points of light on the celestial sphere. Conversely, planets don't twinkle at all because, in essence, they are our celestial roommates. In fact, within our solar system, distances are not measured in light-years but rather in astronomical units (AU). Everything is measured against the distance between Earth and the sun, which is on average 93 million miles (150 million kilometers). This distance equals 1 AU. Mercury, the closest planet to the sun, is 0.3 AU from it. On the other hand, dwarf planet Pluto is 40 AU from the sun.

Courtesy of NASA

On the distance frontier, Mercury and Venus are the "inferior planets" because they orbit the sun in closer proximity than does Earth. Courtesy of their outer-space coordinates, Mercury and Venus are thus observable near the times of sunset or sunrise—and never for an entire night.

The remaining planets that strut their stuff—beyond Earth's orbit—are the "superior planets" in the distance game. On occasion, superior planets are in "opposition." That is, they are directly opposite the sun from our earthly perspective, which supplies us with the stargazing equivalent of a ringside seat.

Mercury

The runt of the planet family, Mercury, touts a diameter of only 3,032 miles (4,879 kilometers). Its atmosphere is considered downright primitive. Its surface is akin to the moon's—extremely cratered and austere in appearance. Mercury orbits the sun faster—in approximately 88 days—than any of the other seven planets in our solar system. Yet, a complete Mercury day takes 58 Earth days. The planet's average distance from the sun during its orbital odyssey is just 36 million miles (58 million kilometers).

Mercury is challenging stargazing prey. For starters, it's both very small and very near the sun. The sun's brilliance reflecting

Courtesy of NASA

off of its bleak surface—without a viable atmosphere to absorb and disseminate the light—further compounds this observational conundrum. Nonetheless, depending on its orbital position, Mercury can be spotted very low on the horizon in the early morning or early evening hours.

When hunting down Mercury, consider this little celestial bit of trivia: From its surface, the sun in the sky would appear about two and a half times the size that we perceive here on Earth. This glaring reality goes a long way toward explaining why Mercury is often a tough night-sky find.

Venus

From our perspective, Venus is often the brightest celestial body in the night sky, excluding the moon but including both stars and planets. Appearing low in the sky in the early evenings or early mornings, depending on its orbital position, it is highly visible to the unaided eye. However, despite its captivating brightness on

© Andre Nantel | Dreamstime.com

the nighttime stage, closer looks at Venus won't supply you with a glimpse of its surface features. This is because Venus is blanketed in multiple cloud layers, which rain down sulfuric acid on the hapless terrain below.

Venus has been called "Earth's twin" because the two planets are relatively near one another and sport similar dimensions, although Earth is somewhat larger. But there the similarities end. Venus is exceptionally hot and bone-dry; there is no water on the planet because of its off-the-chart temperatures. While Venus is nearly twice as far from the sun as Mercury, its surface is considerably hotter.

Venus's orbiting patterns are unique, too. It rotates exceptionally slowly on its axis—so slowly that it completes a rotation around the sun faster than a single axis revolution. A Venus day is thus longer than a Venus year.

Mars

Mars's celestial sphere dimensions and luminosity vary dramatically depending on its orbital position. It is readily spotted with

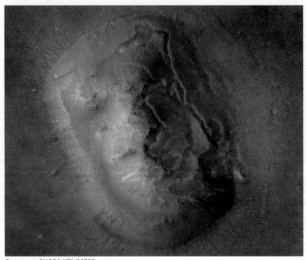

Courtesy of NASA/JPL/MSSS

the naked eye when it is nearest Earth. This close proximity occurs for approximately three to four months each year.

When the so-called Red Planet is ripe for viewing, focus your telescope on this intriguing space body and take a good look around. Exploring its surface is at once worthwhile and even breathtaking on occasion. Under optimal conditions, Mars permits stargazers intimate views of its diverse landscape.

In fact, amateur telescopes can take you places on Mars that no other planet permits. While it's relatively small in size—about half of Earth's dimensions—Mars nonetheless touts taller mountains and deeper valleys than we have here. Telescopes can locate the dark, highly cratered terrain in the planet's southern regions, as well as the lighter-colored and smoother plains up north. When conditions cooperate, Mars's polar ice caps can be seen. There's also the "Face of Mars" to check out, which is visible in the planet's northern hemisphere. Popular culture has made this region of Mars must-see night-sky quarry.

Jupiter

To increase your chances of getting a prime view of Jupiter, explore this gas-giant planet when it is positioned opposite the sun and in line with Earth. During these moments of opposition, Jupiter is nearest the Earth's orbit and appears considerable in size and brightness. While Venus is the brightest object outside of the moon in the night sky, it is not visible in the late evening hours. Jupiter is typically the brightest "star" on the stage throughout the night.

For amateur telescope users, Jupiter is renowned for revealing its atmosphere, which is its surface, in great detail. When zeroing in on this gas giant, what you are seeing are the tops of dense clouds high up in the planet's atmosphere. Jupiter has no solid surface to speak of, but its gaseous innards nonetheless get denser and denser as they get closer and closer to the planet's core.

When conditions allow, even basic telescopes can distinguish the bands of the oval-shaped "Great Red Spot" on Jupiter.

Courtesy of NASA

The Great Red Spot consists of gaseous cloud tops both appreciably higher and colder than neighboring regions of the planet's atmosphere. The Great Red Spot is large enough to accommodate two Earths.

Saturn

The planet Saturn's likeness is recognized by people of all ages in all regions of the world. Courtesy of its lustrous ring structure, it is widely appreciated as an object of sheer beauty in our solar system. So, not surprisingly, Saturn is of particular interest to observers of the night sky.

But while Saturn is a popular solar system inhabitant and can be seen with the naked eye and small telescopes, you will not get a handle on the planet's famous rings. Although the biggest rings surrounding Saturn are exceptionally wide—up to 180,000 miles (300,000 kilometers)—they are also paper-thin,

© Sabino Parente | Dreamstime.com

making them difficult and usually impossible to see from Earth. NASA flyby satellites like *Cassini* have taken the best photographs of Saturn.

What you will likely see when observing this gas giant are thick clouds of gas. These suffocating clouds swathe Saturn, giving it a uniform, almost bland look. Occasionally, the planet exhibits cloud patterns that are something akin to Jupiter's Great Red Spot. Saturn was the outermost planet known to ancient observers of the night sky. Uranus was originally thought to be a star, and Neptune was not discovered until more powerful telescopes made it possible.

Uranus

Uranus is the most distant planet in our solar system that can be seen without a telescope's magnifying brawn. Ancient astronomers noted its existence in the night sky but did not classify it as a planet because of its faint illumination and seemingly deliberate orbit. They considered it just one star among many.

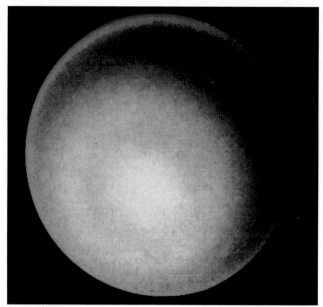

Courtesy of NASA

Uranus's pale blue-green clouds are its defining feature. These omnipresent clouds consist of minuscule methane gas crystals. The planet spins on its axis on the same plane as its orbit around the sun. This atypical positioning and movement supplies Uranus with both unpredictable and very harsh weather conditions.

When all is said and done, Uranus is literally frozen solid. Astronomers often refer to the planet as an "ice giant" to distinguish it from its more prominent gas-giant peers. While Uranus is similar to Jupiter in gas composition, it reveals many more ices, including water, ammonia, and methane. With an interior consisting of frozen gases and rock, it radiates negligible heat in comparison with Jupiter, Saturn, and Neptune. Its atmosphere's temperature is -355° F (-215° C)—cold indeed.

Neptune

The planet Neptune has the distinction of being the only planet that cannot be observed with the naked eye under any circumstances. A telescope is always necessary in order to hunt it down in the outer recesses of the solar system.

Neptune is the eighth—and last—planet from the sun. It is, in fact, thirty times farther from the sun than the Earth is. And whereas our home planet orbits the sun once every year, Neptune takes around 165 years to complete the trip.

The rather consistently bluish bearing of Neptune is derivative of the planet's frozen methane clouds, which absorb other colors. These clouds give the illusion of a solid surface and sometimes sport a "Great Dark Spot," not wholly different from Jupiter's Great Red Spot. The Great Dark Spot resides in the planet's southern hemisphere and is caused by anticyclonic storms. Winds have been measured at 1,500 miles (2,400 kilometers) per hour in this area. The Great Dark Spot reveals itself in darker shades of blue. However, unlike the Great Red Spot of Jupiter, which lasts decades, the Great Dark Spot of Neptune comes and goes every few years.

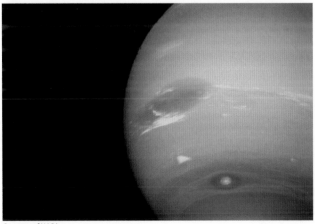

Courtesy of NASA

CHAPTER 10 Exploring the Deep Sky

Galaxies

Amateur astronomers frequently employ the Charles Messier playbook while skywatching—and for good reason. The eighteenth-century astronomer observed and cataloged 110 deep-sky objects, including galaxies. But of all the objects chronicled, galaxies are without question the most frustrating celestial game to bag. This is because they are typically very far away and suffused in an amalgam of billions of stars' light.

However, it is this characteristic brightness that permits many galaxy sightings with binoculars and small telescopes.

Courtesy of NASA

Courtesy of NASA

When galaxy hunting, it is vital for the night sky to be as clear and as dark as possible—free of excessive moonlight and artificial light sources on the ground.

Approximately 77 percent of the observed galaxies in the universe are classified as spiral galaxies. Elliptical galaxies, which are ellipsoidal in shape and display no spiral arms to speak of, are the next most common galaxy type. Elliptical galaxies generally accommodate fewer stars than spiral galaxies and tend to reside in clusters of galaxies. Only 3 percent of observed galaxies cannot be labeled spiral or elliptical. They are called "irregular." There are also subclassifications in these three broad categories.

The very best pictures of galaxies are often taken in conjunction with powerful NASA telescopes. The impressive barred spiral galaxy NGC 1300, for instance, is a difficult find for hobby amateurs. At the very least, a mid-level telescope would be required to locate this celestial body in Eridanus, and even then, ascertaining the galaxy's distinctive contours would be almost impossible.

Fortunately, there are galaxies that are accessible stargazing targets, including M31 and M32. Akin to our Milky Way, the Andromeda Galaxy, M31, contains hundreds of billions of stars bound together by very persuasive gravitational forces. Our neighboring galaxy is some 2.2 million light-years away, with a satellite dwarf elliptical galaxy, M32, close at hand. M32 is small in size but an unusually bright companion worth hunting down in the Andromeda constellation.

Searching, too, for M77, the barred spiral galaxy in the constellation Cetus, is potentially a fruitful undertaking. Both M81 and M82 in the northern constellation Ursa Major are very bright and easily spotted. M82, by the way, is sometimes referred to as a "starburst galaxy" because it is spawning new stars at a breathtaking pace. Ursa Major, of course, is home to the Big Dipper.

Nebulae

Nebulae in the cosmos are another category of deep-sky objects that engage the interest of professionals and hobbyists alike. These gaseous and dust-riddled regions of space are often both

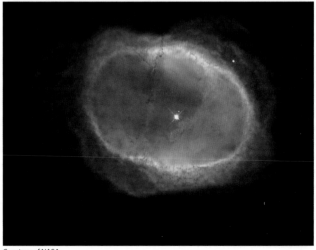

Courtesy of NASA

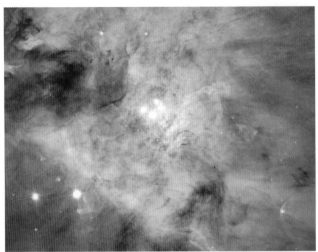

Courtesy of NASA [http://www.nasa.gov/]. ESA [http://spacetelescope.org], M. Robberto (Space Telescope Science Institute [http://www.stsci.edu]/ESA[http://www.spacetelescope.org], and the Hubble Space Telescope Orion Treasury Project Team

vibrant and infused with colors. The "planetary nebula," a particular classification, is the derivative of a dying star not unlike our sun. Planetary nebulae have nothing to do with planets.

The Ring Nebula is a prime example of one such star's colorful transition from red giant star to white dwarf star. This celestial body, sporting shades of blue, yellow-green, and red, is a prime catch for an amateur astronomer. The blue is ionized helium; the yellow-green, doubly ionized oxygen; and the red, ionized nitrogen. The Ring Nebula is considered a quintessential planetary nebula because of its very defined shape and multi-hued gases. Having expelled its outer layers into the celestial ether, the dying star at the nebula's center supplies its radiant nebulosity.

The Ring Nebula is easy enough to spot. For starters, locate the bright star Vega in Lyra; the Ring Nebula is due south of this prominent star, which is part of the Summer Triangle. And, naturally, summertime is the best time to check it out.

Nebulae like the Orion Nebula, M42, are categorized as "diffuse" or "emission" nebulae, and are decidedly more sprawling than their planetary nebulae brethren. Within these celestial clouds of dust and gas are stellar nurseries, breeding grounds for new star formation. Gravity's imposing and unrelenting tugs-of-war condense gas and dust into smaller, more compact masses that become the stars of tomorrow.

The Orion Nebula can, in fact, be seen with the naked eye due south of the mighty hunter's renowned belt of stars. However, without binoculars or a telescope, its nebulosity cannot be deciphered. With these visual aids, however, its gaseousness comes to life and can be fully appreciated.

The gases amid the Orion Nebula, believed to be mostly hydrogen, emit a radiant glow that sheaths a mother lode of hot young stars. This potent one-two punch of gaseous clouds and newly forming stars within the interstellar medium is the perfect recipe for outer-space illumination. At a relatively close 1,500 light-years from Earth, the Orion Nebula furnishes you with a rare insider's view into stellar evolution. Winter is a prime time to explore this multihued celestial cloud of gas and dust.

© a. v. ley | Shutterstock

Like the Orion Nebula, the Lagoon Nebula, M8, is an emission nebula. This ample interstellar cloud is located in the zodiac constellation Sagittarius. It looms north of the Sagittarius star cloud, or just west of the top star in the familiar Teapot asterism in the constellation's eastern quadrant. However, the Lagoon Nebula is barely visible to the naked eye, and then, only under optimal conditions.

Fortunately, observation with binoculars or telescope brings it nicely into focus. Interestingly, the nebula appears predominantly gray when observed through these magnifying devices, but red and pinkish in time-exposure color photographs. This popular starwatching target nonetheless reveals a well-defined oval form, which you will discern upon observation. You will have no trouble pinpointing its well-lit center, too. It is precisely here that astronomers believe life is being breathed into future stars. A darkened pathway also runs through the nebula. It is this winding celestial corridor that inspired the Lagoon Nebula's appropriate appellation.

The region of the night sky that hosts the Lagoon Nebula is among the brightest and most fertile star-forming locales. This very active emission nebula has even fathered its own cluster of hatchling stars: NGC 6530.

Star Clusters

Stars assume their celestial addresses not by mere chance but for a whole host of astronomical reasons, including the most compelling of all: the forces of gravity. Most stars originate in groupings. Like our sun, some of them end up standing alone, but many others form binary systems or multiple systems of stars that orbit one another. There are also clusters of stars that understandably attract the interest of the starstruck.

Open clusters, sometimes called "galactic clusters," consist of rather dispersed youthful stars. They are typically found in the environs of the Milky Way's plane. Contrarily, globular clusters are dense groupings of considerably older stars—sometimes hosting millions of residents—that form tightly wound spheres. Most

© Diego Barucco | Dreamstime.com

of the globular clusters in our galaxy are very near its searing, metal-rich nucleus, and some of them are difficult to detect because of their celestial coordinates, which are often shrouded in opaque space dust.

The brimming constellation Sagittarius, which spans this region of space, hosts many of these globular clusters, including the very visible M22. A prominent open cluster, M8, is also found in this zodiac constellation, which is best explored in the summertime.

In addition to the Pleiades Cluster, Double Cluster in Perseus, and the Great Globular Cluster, there are many other star clusters worth seeking out. In Scorpius the Scorpion, which inches along the summer's southern sky, multiple star clusters exist. The Butterfly Cluster (M6) and the Ptolemy Cluster (M7) are two very accessible open clusters. Globular cluster M4 is one of the closest to Earth—and one of the most impressive, too—with globular cluster M80 nearby.

Auriga the Charioteer, due north of Orion, greatly benefits from the cluster-rich Milky Way that winds through it. Three bright open clusters can be readily observed with binoculars: M36, M37, and M38. However, a telescope's magnifying might is required to get a good fix on the individual stars in the clusters.

Other discernible open clusters include M35 in Gemini, M45 in Taurus, and NGC 457 in Cassiopeia. The somewhat-obscure southern constellation Puppis reaps the rewards of the Milky Way calling on its celestial turf, and touts two rich open clusters, M46 and M47. Globular clusters to consider adding to your stargazing menu include M3 in Canes Venatici, M92 in Hercules, M12 in Ophiuchus, M53 in Coma Berenices, and M71 in Sagitta.

Courtesy of Atlas Image obtained as part of the Two Micron All Sky Survey (2MASS), a joint project of the University of Massachusetts and the Infrared Processing and Analysis Center/California Institute of Technology, funded by the National Aeronautics and Space Administration and the National Science Foundation

CHAPTER 11

Night-Sky Phenomena
and Visitations

Meteors

Smaller than asteroids but bigger than atoms, meteoroids are solid space debris. These fragments of dust, metals, and rock whizzing through space in chaotic orbits around the sun, or other celestial bodies, regularly enter Earth's orbit. Meteors can be seen at just about any time, although there are specific times of the year when prolific meteor showers occur.

Foremost, remember that meteor streaks are often both muted and fast-moving, so if you want to increase your chances of seeing meteors in the night sky, or annual meteor showers at their most prolific, it's a good idea to be as far away from man-made

© Dnally | Dreamstime.com

Illustration by David Cole Wheeler

lighting sources as possible, including city lights, automobile traffic, etc. Illuminated distractions here on the ground both reduce one's very sensitive night vision and make faint celestial objects and phenomena even fainter. The night sky also needs to be especially clear and not dominated by excessive moonlight.

If you make "secluded, dark, and safe" your credo during all astronomical safaris, you'll see plenty, including both meteors and the always-impressive meteor showers. Since meteors are at once unpredictable and short-lived, naked-eye observation works well. However, many weaker meteors and meteor showers can be captured with binoculars and telescopes.

Comets and asteroids orbiting the sun are simultaneously leaving trails of dusty and rocky debris in their wakes. When Earth's conventional orbit intersects their beaten paths—at around the same time every year—meteor showers occur.

The celestial fireworks known as the Quadrantids do their thing from around December 28 to January 7 every year in the vicinity of the constellation Bootes. Between April 16 and April 26, the April Lyrids can be seen at their most fruitful in the constellation Lyra.

The summertime Perseids, radiating from the constellation Perseus the Hero, reveal themselves in all their splendor from mid-July through most of the month of August. The Orionids occur in the constellation Orion from mid- to late October. Mid-November ushers in perhaps the most prolific of all meteor showers: the Leonids. The region to focus on is the constellation Leo, where the Leonids streak across the sky in great numbers. Last but not least, the Geminids, showering down across the borders of the constellation Gemini, spring onto the scene in mid-December, with December 12 through 14 a typical peak time.

Aurora Borealis

The layers of exceptionally hot gases that comprise our sun incessantly send solar flares careening through the corridors of interplanetary space. In the coming years, scientists anticipate ever-more-dramatic magnetic energy being unleashed from the solar atmosphere. This is good news for aurorae hunters—for those

© Jens Mayer | Dreamstime.com

interested in witnessing the spectacular space weather otherwise known as the Aurora Borealis, or Northern Lights, in the Northern Hemisphere.

Solar radiation from outer space meeting and greeting the myriad gases contained in our upper atmosphere—approximately one-third of the distance to the moon—inspire these cascading light shows. Oxygen atoms, for instance, are wont to produce red colors at the highest altitudes. From our vantage point, the red often appears as a brownish-red. Red aurorae are also the rarest. At lower altitudes, the same oxygen gas unleashes greens, yellows, and oranges. The presence of nitrogen frequently delivers colors like blue and purple.

These myriad colors associated with aurorae—red, green, yellow, blue, orange, and purple—make these atmospheric light shows worth the price of admission. Aurorae are never the same in intensity or color schemes. Some appear pale and short-lived, whereas others are vivid and carpet the night sky.

The odds of witnessing the Aurora Borealis increase as you near the North Pole. Optimally, you need to be within a prescribed oval band—one with its epicenter at the North Magnetic Pole, and with a width that ranges from 6 to 600 miles (10 to 1,000

© Jostein Hauge | Dreamstime.com

kilometers). Being inside or close to this band places you on prime terra firma to experience nature's greatest light show.

This oval, however, does not conform to uniform latitudes. For example, East Coast denizens of the United States are more apt to behold the Aurora Borealis than folks on the West Coast. And during solar maximum activity, aurorae have been known to travel far afield of this band.

It should also be pointed out that locations above the frigid Arctic Circle are bathed in sunlight 24/7 from April through September—spring and summer in the Northern Hemisphere. And since the Northern Lights are predominantly nighttime events—although twilight aurorae are known to impress— wintertime supplies the most dazzling performances in these places. Generally speaking, the liveliest and most impressive light shows occur near the midnight hour. Peak aurorae viewing hours are between 11 p.m. and 2 a.m.

Comets

Comets have an unmatched celestial aura. They aren't stars, planets, moons, or nebulae, which are always where we expect to find them, but wholly unique space objects instead, with—in cosmological terms—short life spans. Comets are a blend of frozen gases, dust, and solids. When they lose their volatility over time, via melting and evaporating gases, they morph into ordinary slabs of rock meandering through the ether. Scientists believe that many asteroids are, in fact, former comet nuclei.

Comets travel in sweeping elliptical orbits around the sun, which takes them both very deep into our solar system and very near the sun. Their contours are highly irregular; no two are the same. Amateur astronomers have discovered countless comets through the years. However, the Central Bureau for Astronomical Telegrams (CBAT), a branch of the International Astronomical Union (IAU), says that for every legitimate comet sighting reported to it, five reports amount to nothing.

The IAU implores skywatchers to, first and foremost, make sure that what they are seeing is the real McCoy. Look for motion.

© Wikipedia Commons/NASA/W. Liller/ NSSDC's Photo Gallery

Comets are always on the move, never idle. It's easy to be fooled by space's ghostly images and optical illusions.

The scientific community classifies comets as either short-term or long-term. Short-term comets require 200 years or less to orbit the sun. Long-term comets need more time than that— thousands of years sometimes. Halley's Comet is a short-term comet. Comets Hyakutake and Hale-Bopp are both long-term.

A comet's head and atmosphere typically consist of an expanding and fuzzy cloud of widely scattered materials called its "coma." The comet's engine—a small but nonetheless radiant and solid nucleus—lies at the coma's center and will normally swell in size and brightness as it nears the sun's extreme heat and robust solar wind. Also, as the comet ventures closer to the sun, its coma will leave visible tracks of gas and dust in its wake. When illuminated by the sun, this cosmic residue—the comet's "tail," as it were—is a breathtaking sight.

© Martin Ezequiel Gardeazabal | Shutterstock

There are untold numbers of comets in outer space, and the vast majority of them are too minuscule or faint to be observed without the aid of a telescope. The comets that reveal themselves to us—sometimes in grand fashion—are always passing very near the sun.

Index

Index

About the Author

Nicholas Nigro is a freelance writer and the author of several books in a variety of fields. From popular culture to business, pets and animals to science and medicine, his publishing credits span a broad spectrum of topics and appeal to an eclectic swath of readers. A skywatcher always, and a backyard astronomer when time and circumstances permit, his most recent book, *KNACK Night Sky,* is a colorful ode to our dynamic solar system and the sprawling and enigmatic universe beyond. He lives in New York City but stargazes elsewhere.